THE COMET OF 44 B.C. AND CAESAR'S FUNERAL GAMES

AMERICAN PHILOLOGICAL ASSOCIATION
American Classical Studies

Series Editor

David L. Blank

Number 39

THE COMET OF 44 B.C. AND CAESAR'S FUNERAL GAMES

by
John T. Ramsey and A. Lewis Licht

John T. Ramsey

Department of Classics
The University of Illinois at Chicago

and

A. Lewis Licht

Department of Physics
The University of Illinois at Chicago

THE COMET OF 44 B.C. AND CAESAR'S FUNERAL GAMES

Foreword by
Brian G. Marsden
Harvard-Smithsonian Center
for Astrophysics

Scholars Press
Atlanta, Georgia

THE COMET OF 44 B.C. AND CAESAR'S FUNERAL GAMES

by
John T. Ramsey
and
A. Lewis Licht

© 1997
The American Philological Association

Set in Times Roman by the Office of Publications Services
The University of Illinois at Chicago

Library of Congress Cataloging in Publication Data
Ramsey, J. T. (John T.)
    The comet of 44 B.C. and Caesar's funeral games / John T.
Ramsey and A. Lewis Licht.
      p.   cm. — (American classical studies ; no. 39)
    Includes bibliographical references.
    ISBN 0-7885-0273-5 (cloth : alk. paper). — ISBN 0-7885-0274-3
(paper : alk. paper)
    1. Festivals—Rome—History—Sources. 2. Comets—44 B.C.—
Sources. 3. Caesar, Julius—Cults. 4. Augustus, Emperor of Rome,
63 B.C.–14 A.D. 5. Comets—Religious aspects.   I. Licht, A. Lewis.
II. Title. III. Series.
GT4852.R6R36   1996
394.2'6937—dc20   96-17535
                                        CIP

Printed in the United States of America
on acid-free paper

For Sarah and Sheila

CONTENTS

ILLUSTRATIONS

FOREWORD

"When beggars die, there are no comets seen;
The heavens themselves blaze forth the death of princes."

These lines, attributed to Calpurnia on the morning of March 15, 44 B.C., as she implored her husband not to go into the office that fateful day, are arguably the most famous of all literary allusions to comets. In writing them, Shakespeare was not only expressing the traditional association between disasters on the earth and the appearance of portents in the sky, but he was also demonstrating his appreciation of the historical account of the specific comet that allegedly appeared soon after the assassination of Julius Caesar.

It is the proper documentation of Caesar's Comet that is the principal concern of this monograph. At its basic level, that documentation is the domain of the classicist. John Ramsey demonstrates his scholarship in this area by bringing together and discussing the numerous historical sources on the events in and around Rome in 44 B.C. and surrounding years. But that cannot be the whole story. Comets are not "local" phenomena. A full treatment of Caesar's comet must attempt to correlate the Roman information with what one may or may not be able to learn of observations of it from other parts of the world, and how observations of a single comet can tie in with what we know of other comets and about the cometary phenomenon as a whole. This is the domain of the astronomer or physicist. Here Lewis Licht enters with his application of the scientific method in an ingenious attempt to augment considerably what was previously known about the reaction of the skies to Caesar's death. Whether one believes this argument or not, the outcome of this interdisciplinary study is an intriguing account and an excellent illustration of the importance of such collaborations in producing something larger than would be possible for a classicist or a physicist alone.

The fundamental problem, of course, is that science in Rome of 44 B.C. had not developed to the point that records were kept of the configurations of bright comets relative to the stars among which they moved from night to night. On the contrary, essentially every one of the historical records is based on a single description by Caesar's adopted son Octavian. Worse still, Octavian did not set that description down for 20 years when, as Augustus Caesar, he was writing his memoirs and casually mentioned that a comet appeared during the Games he held in honor of his predecessor.

Such vagueness was a hallmark of all the early European records of comets. Indeed, just three centuries ago, when Halley made his celebrated first computations of cometary orbits on the basis of gravitational theory, he bemoaned the fact that "I find nothing at all that can be of service in this affair before A.D. 1337". However, this did not stop him from speculating on the identities of comets observed in earlier years, even if his suggestion that the comet of 1305 was an earlier manifestation of his celebrated 75-year-period comet was subsequently shown to be erroneous.

In a more extreme speculation, Halley attempted to identify the great comet of 1680 with Caesar's Comet and others that seemed to fit in with a revolution period of 575 years. In so doing, Halley followed Gruter in supposing that the Games began on September 23. He computed that, if the comet had been at its closest to the sun on September 18, it would have risen during the Games at the eleventh hour of the night and been located under the "septem triones", or seven bright stars of Ursa Major (i.e., the Big Dipper or the Plough, according as to whether you are reading this in the United States or the United Kingdom). For the time it was made, Halley's conjecture had considerable merit, although Octavian's actual statement is that the comet rose during the eleventh hour of the day, rather than the night—in which case one would have to interpret "septem triones", or "septentriones", in the more general sense as indicating the northern part of the sky. Furthermore, careful computations later by Encke render it very unlikely that the period of the 1680 comet could be as short as 575 years, and I am not alone in tending instead to equate the comet of 1106—an important element in Halley's scenario—with the "Kreutz group", specifically perhaps as the joint parent of the sungrazing comets of 1882 and 1965. Now, Ramsey and Licht conclusively show that the Games in 44 B.C. were held in late July, rather than in late September, another departure from what Halley proposed. Correcting this error, which was copied by the cometographer Pingré and others, may be the single most important element of the present work on the astronomical side.

Although European sources of cometary observations have little value before the fourteenth century (and even the fifteenth), there are more useful records of many ancient comets from further east, notably from China. Thanks to the travels of Melchisedech Thévenot, a few of those records were already known in Europe in Halley's time, but it was not until later in the eighteenth century that information became available concerning Chinese cometary observations made around the time of Caesar. That information, mainly from Mailla's French translation of a general history of China and from manuscripts by Antoine Gaubil, was correlated by Pingré with the European data. The Chinese reports, confirmed by subsequent researchers as discussed in the present work, make it quite clear that the only comet recorded in the extant Chinese sources within several years of Caesar's death did indeed appear in 44 B.C., but as early as May-June. Pingré, who clearly accepted Halley's hypothesis of the identity of the Roman comet with that of 1680, yet appreciated the inconsistency with the Chinese report, had no choice but to suppose that two separate comets had been observed in 44 B.C.

By changing the date of Caesar's Comet to July, and at the same time abandoning the 1680 identity, the possibility that the Roman and Chinese comets were one and the same gains merit. The second main thesis of the present work is the attempt to reinforce this argument. In so doing, the authors present a novel method of orbit determination, and I have verified the use of this method with alternative computations of my own. Of course, there is the problem of why the Romans did not report the comet in May and why the Chinese did not report it in July, but this holds whether there were two comets or one. In either case, reasons why this might be so are offered.

Much more troublesome is that the Roman comet was evidently bright enough to be seen in broad daylight. Given its presence in the north and thus its apparent near circumpolarity, it is difficult to account for the absence of reports of it during the night-time, when it would surely have been even more spectacular. If the May-June and July comets were identical, there is also the need to explain why it was so very bright in July, given that the orbit computations then put the comet much farther from both the sun and the earth than in May-June. Again, the authors offer possible reasons for this. Altogether, then, they are able to put forward a scenario that is every bit as logical as Halley's and compatible with facts available to them and not to him.

Did the Roman report really refer to a comet? There is no mention of a tail to distinguish it from a supernova, say. But a supernova hypothesis again suffers from the problems of the lack of observations at night and from China, given also that such a phenomenon could be expected to be more long-lived than a comet. Is the Roman observation entirely fictitious? Given the circumstance of a single reporter two decades after the event, I should be remiss if I were not to consider this as a serious possibility. The whole scientific problem then disappears, but this would be deeply disturbing to the classical community. So it's up to you, dear reader. Whatever you believe, the tale of the comet(s) of 44 B.C. is a fine piece of interdisciplinary detective work and makes fascinating reading.

Brian G. Marsden
Harvard-Smithsonian Center for Astrophysics
Cambridge, MA 02138
December 1996

PREFACE

This work is the product of a unique collaboration between a classicist and a physicist that began in 1993. It is our hope that the melding of these two fields will help to further our understanding of the spectacular, daylight comet that was observed in 44 B.C. during the games that the future emperor Augustus gave in honor of the late Julius Caesar. That comet possesses great historical significance because it came to be interpreted as a sign of Caesar's apotheosis, and it has been celebrated and written about over the course of many centuries. The Chinese too observed what was most likely this same comet and had their own view of its significance. We are fortunate to have this independent evidence of a sighting in 44, untainted by any suspicion that it could have been invented merely to lend greater historical significance to Caesar's murder on the Ides (15th) of March. In Italy, the comet caught the fancy, it seems, of contemporary astrologers and prophets, and centuries later it intrigued the scientific minds of Sir Isaac Newton and Sir Edmund Halley.

To do justice to this multifaceted topic, our study must range well beyond the traditional bounds of classical scholarship, and keeping in mind the breadth of the audience for which this work is intended, we have tried to make it accessible to readers and scholars in many fields besides our own. We have translated into English all texts quoted in Greek or Latin; with the help of colleagues, we have consulted the relevant Chinese and Korean sources in the original and have provided fresh, accurate translations rather than relying upon older and often misleading English versions of those texts; and we have written précis to introduce sections of the discussion that have a technical, scientific nature. We have also taken great care to assemble, translate, and annotate in Appendix I what is intended to be the most comprehensive collection to date of Greco-Roman sources having to do with the comet of 44, its interpretation, and later fame (36 texts in all: nos. 1-33 + 1A, 3A, and 11A). Those texts are printed with facing English translation and are referred to within the body of the work, and in the footnotes, by their numerical designation in the Appendix preceded by the word "Text".

As the title of this book indicates, we discuss both the comet and the historical context in which it appeared: a Roman festival to which funeral games for Julius Caesar were added in 44 B.C. The funeral games were, of course, held just the one time, in the year of Caesar's murder, but the Roman public festival to which the funeral games were attached continued to be celebrated annually on 20-30 July, and in imperial times the festival was known as the *ludi Victoriae Caesaris*. The prevailing modern view holds that the *ludi Victoriae Caesaris* were created to honor Caesar in his lifetime and were presented for the first time in July 45 B.C. We hope to show, on the contrary, that 44 B.C. was the first year in which games honoring Caesar were held in July, and we shall also argue that in 44 the public festival had not yet shed its roots which lay in a set of games that Caesar had founded in 46 B.C. in honor of the goddess Venus Genetrix. Since our ancient sources refer to the festival in 44 B.C. both by its old name (*ludi Veneris Genetricis*) and by its new one (*ludi Victoriae Caesaris*), we have deliberately selected and incorporated into our title the one element that was unquestionably part of the celebration in that year. The expression "Caesar's Funeral Games" has the further advantage of being capable of being understood in two different and equally appropriate senses, as games given by a Caesar (Octavian: the future emperor Augustus and adopted son of Julius Caesar) in honor of a Caesar (the late dictator Julius Caesar). By adopting this title, we can satisfactorily convey the historical context of Caesar's comet and leave open for the time being the question whether the spectators in 44 B.C. would have called the public festival, to which the funeral games were attached, the *ludi Veneris Genetricis* or the *ludi Victoriae Caesaris*.

Some of the topics covered in this study were presented by J. Ramsey at a faculty seminar at the University of Illinois at Chicago in November 1993 and also in a paper delivered in December at the Annual Meeting of the American Philological Association in Washington, D.C. We gratefully acknowledge the advice and encouragement offered on those two occasions by friends and colleagues.

In addition to the acknowledgements that are found in the notes, we wish to express our gratitude to the following:

We thank the Adler Planetarium in Chicago and its staff for its kind cooperation in responding to our questions and for permitting us to consult its rich collection of early books on astronomy. In particular we thank Dr. Eric Carlson, astronomer; Dr. Evelyn Natividad, library administrator; Dr. Marvin Bolt, assistant curator of the History of Astronomy Department; and Mave Lawler also of that department. We are indebted to Nancy Romero, cataloguer of the rare book collection at the University of Illinois at Urbana-Champaign, for a photocopy of the 1715 edition of Sir Edmund Halley's *Synopsis of the Astronomy of Comets*. Dr. Tai Wen Pai, Chinese librarian of the East Asian collection in the Regenstein Library at the University of Chicago, very kindly helped us secure copies of all the Chinese texts that attest the comet of 44 B.C. At our own institution, Kathleen Kilian, the interlibrary loan librarian, provided invaluable assistance by securing copies of difficult to find monographs and articles relevant to our study.

We have benefitted greatly from an extensive correspondence and quite a few conversations over more than three years with Dr. Brian Marsden of the Smithsonian Astrophysical Observatory in Cambridge, Massachusetts. In September 1996, Dr. Marsden very kindly read our manuscript, and his comments have led to the introduction of numerous improvements and new ideas.

We are grateful to the following scholars for reading and commenting on earlier drafts of parts, or the whole of this study: Profs. G. W. Bowersock, G. P. Goold, Alexander Jones, David Pingree, D. R. Shackleton Bailey, F. Richard Stephenson, and Gary Kronk. We are grateful as well to the two readers for the APA who commented on our submission: Prof. Robert Gurval and an anonymous referee. In response to their suggestions and queries, we now devote greater space to such topics as the reliability of the Chinese sources that attest the comet of 44 B.C., whether there was one comet or two in that year, and why it is that Cicero makes no mention of the *sidus Iulium* (a silence that is bound to raise questions). We also assess the potentially suspicious historical tradition that not only a comet but also an eruption of a volcano (most likely Etna) occurred not many months after the portentous Ides of March. For guidance in evaluating the exciting new evidence from the GISP2 ice core in Greenland that appears to confirm our literary sources that report an eruption of Mt. Etna in 44, we thank Prof. Gregory Zielinski of the Glacier Research Group of the University of New Hampshire.

Our colleague John Rohsenow in Linguistics kindly read through the sections of our manuscript dealing with the Chinese sources and their interpretation, vetting those pages for consistency in adhering to the conventions of Romanization. Members of the Needham Research Institute in Cambridge, England, assisted on numerous occasions in connection with issues having to do with the East Asian sources. In particular we thank Dr. Ho Peng Yoke, Director; Dr. Christopher Cullen, Deputy Director; and Mr. John Moffett, librarian; all of whom corresponded with us.

Shortly before his untimely death, Prof. Pierre Brind'Amour very kindly reviewed the parts of our argument having to do with the Roman calendar. He also provided us with his splendid *SkyClock* v. 1.1 for MS.DOS (*SKCL*), comprising two programs copyrighted 1988: "Clock" for converting dates to a host of different calendars and "Sky", an electronic ephemeris for calculating the position of the Sun, Moon, planets, and stars as seen from any location on Earth, at any given hour, in any given year. We used "Sky" to produce the data in Table 2 and to verify some of the calculations that we had arrived at by means of a program written by L. Licht.

Of course, none of the above-mentioned scholars who advised us and commented on our work is to be held responsible for any of the views expressed here.

Ramsey's work was supported by a Fellowship for College Teachers, awarded by the National Endowment for the Humanities for the year 1993-94, and by a grant for the purchase of word-processing equipment from the Campus Research Board of the University of Illinois at Chicago. We thank the Dean of the College of Liberal Arts and Sciences, the Campus Research Board, and the Department of Physics for providing funds for typesetting this book. We thank Tim Northrup of UIC Publications Services for typesetting the manuscript and our colleague Alexander MacGregor for his unstinting help with the page proofs. Last, but by no means least, we express our special thanks to the editor of this series, Prof. David Blank, for his constant support and encouragement of our work. This book could not have been brought to fruition without his expert guidance.

J.T.R.
A.L.L.

Chicago
December 1996

ABBREVIATIONS

Works by Cicero are generally cited by title only; references to Cicero's letters by the numbers assigned by Shackleton Bailey are according to his Cambridge editions. References to Appian are to his *Bella Civilia,* and references to Nicolaus of Damascus (Nic. Dam.) are to his biography of Augustus and related texts (frags. 125-130, ed. Jacoby). Cross references and entries in the indices having the form p. 115.61 are to be understood as p. 115 n. 61. The names of ancient authors and their works are abbreviated as indicated in our Index of Sources, and the titles of classical journals are abbreviated as in *L' Année Philologique.* In addition, standard collections and works of reference are abbreviated as follows:

CIL	*Corpus Inscriptionum Latinarum* (Berlin: Reimer, 1863—)
Dar.-Sag.	Daremberg, Ch. and E. Saglio, *Dictionnaire des Antiquités Grecques et Romaines d'après les Textes et les Monuments* (Paris: Hachette, 1877-1919)
HRR	Hermann Peter, *Historicorum Romanorum Reliquiae,* 2 vols. (Leipzig: Teubner, 1867)
HS	*Ch'ien Han-shu* (History of the Former [Western] Han Dynasty, 206 B.C.-A.D. 9) composed chiefly by Pan Ku (A.D. 32-92) in the first century A.D.
ILLRP	A. Degrassi, *Inscriptiones Latinae Liberae Rei Publicae,* 2 vols. (Florence: La Nova Italia, 1957, 1963)
ILS	H. Dessau, *Inscriptiones Latinae Selectae,* 3 vols. (Berlin: Weidmann, 1892 -1916)
Inscr. Ital.	A. Degrassi, *Inscriptiones Italiae* XIII.1-2 (Rome: La Libreria dello Stato, 1947, 1963)
LSJ	Henry Liddell and Robert Scott, *Greek-English Lexicon,* 9th rev. ed. by Henry Jones (Oxford: Oxford Univ. Pr., 1940)
MRR	T. R. S. Broughton, *Magistrates of the Roman Republic,* rev. ed., 3

	vols. (Atlanta: Scholars Pr., 1952-86)
OCT	Oxford Classical Texts, Oxford University Press
OLD	*Oxford Latin Dictionary,* ed. P. G. W. Glare, (Oxford: Oxford Univ. Pr., 1968-82)
RE	F. Pauly and G. Wissowa, eds. *Real-Encyclopädie der klassischen Altertumswissenschaft* (Stuttgart: Metzler/Alfred Druckenmüller, 1893-1980)
RIC²	C.H.V. Sutherland and R. Carson, *Roman Imperial Coinage.* I, rev. ed. (London: Spink, 1984): numbers designate coins by their catalogue number
SB	D. R. Shackleton Bailey
SKCL	*SkyClock* v. 1.1 for MS.DOS, an electronic ephemeris and calendar conversion program written by Pierre Brind'Amour (1988)
TLL	*Thesaurus Linguae Latinae* (Leipzig: Teubner, 1900—)

I

INTRODUCTION

The Historical Context

When Julius Caesar's great-nephew Gaius Octavius returned to Rome less than two months after Caesar's murder on the Ides of March 44 B.C., he immediately took the legal action needed to accept his inheritance from the late dictator, who had named Octavius his chief heir and adopted son. Octavian, as we may conveniently call this youth of eighteen after he had become "Gaius Julius Caesar Octavianus" by testamentary adoption, formally accepted the terms of Caesar's will by appearing before the acting urban praetor, Gaius Antonius (Appian 3.14.49). One of the tribunes then permitted Octavian to address a public meeting (*contio*) at which he promised to pay the legacies left by Caesar to the Roman people and vowed to hold games.[1] These games were celebrated in due course, and it was during them that a spectacular, daylight comet appeared, causing a belief to spread among the Roman people that Caesar

---

[1] Becht 56-57 assigns the *contio* to 8 or 9 May, citing *Att.* 14.20.5 and 21.4 (both written on 11 May) according to which Cicero was still awaiting news of Octavian's speech: the latest letter that Cicero had received from Atticus as of 11 May had been dispatched from Lanuvium on 9 May (*Att.* 14.20.1). The *contio* took place soon after Octavian's arrival, but we do not know the precise date of his entry into Rome. Yavetz (1969) 73 assumes that Octavian may not have arrived until 11 May, and Weinstock 369 puts Octavian's announcement of the games ca. 18 May, which is the *terminus post quem non* for the *contio* since Cicero (near Sinuessa, in southern Latium) received Atticus' letter describing the *contio* on the evening of the 18th (*Att.* 15.2.3). The tribune who convened the *contio* for Octavian was probably Mark Antony's brother Lucius (*Att.* 14.20.5)—not Ti. Cannutius, so Dio 45.6.3, who confuses this incident with a later one in October—showing that at first the Antonii apparently did not view Octavian as a great threat.

1

had been taken up to heaven where he joined the ranks of the gods.[2] This event had enormous consequences both for the career of Octavian and for the movement that was trying to secure the recognition of the late dictator as a new god. Both that movement and the career of Caesar's young heir were stalled at the time. Mark Antony, the Roman consul and Caesar's would-be political heir, stood in the way.[3] However, after the games and appearance of the comet, Antony was forced to pay greater respect to his eighteen-year-old rival, and he also found it expedient to devise new honors for Caesar. At stake was the leadership of Caesar's followers, especially his military veterans and the urban mob. A public reconciliation was held between Octavian and Antony on the Capitoline,[4] and at a meeting of the Senate on 1 September Antony proposed the addition of a day in Caesar's honor to all Thanksgivings (*supplicationes*).[5]

On these points nearly all of our ancient sources agree. However, closer inspection of these sources reveals a detail that is very curious indeed, one that decidedly conflicts with the prevailing modern view of the nature of the festival in 44 at which the comet appeared. Not only do we find that our Greek and Latin texts call Octavian's games by two different names (*ludi Veneris Genetricis* and *ludi Victoriae Caesaris*), but what is more surprising still, the bulk of our sources, including most likely Octavian himself in his

---

[2] See Texts 1-13 in the Appendix I, esp. 1, 3, 5, 6, 8, and 9 for the apotheosis.

[3] Antony had, for instance, crushed the movement led by Pseudo-Marius in early April that tried to establish worship of Caesar at an altar erected in the Forum on the site of Caesar's cremation (Cic. *Phil.* 1.5, cf. *Att.* 14.8.1; Appian 3.2.3-3.9). Antony had also successfully blocked two attempts of Octavian to display at the public games the gilded chair and crown of Caesar (symbols of his divinity: Weinstock 281-84): Appian 3.28.105-106 and Text 15; Plut. *Ant.* 16.5.

[4] Nic. Dam. 29.115-19; Plut. *Ant.* 16.3; Appian 3.29.111-115; Dio 45.8.1-2. Appian (3.39.156) alone reports a second reconciliation that supposedly took place at the end of September, clearly a doublet of the first: see Ehrenwirth 62-63. The popular demonstrations in Octavian's favor at the games were soon followed, on 1 August, by a verbal attack on Mark Antony in the Senate by Lucius Calpurnius Piso, the father of Caesar's widow Calpurnia (*Att.* 16.7.5, 7; *Phil.* 1.10, 14; 5.19; *Fam.* 12.2.1). These events revealed that Antony could not count upon the unswerving loyalty of Caesar's followers: see Levi I 108-10.

[5] Cic. *Phil.* 1.12-13, 2.110. Dio (45.7.2) mistakenly states that this honor had been passed earlier but had not been put into force; perhaps he confuses it with the measure passed in Caesar's lifetime that added a day in his honor to all triumphs (Dio 43.44.6). Dio also incorrectly implies that no use had been made of the new name "July" for the month Quintilis. While it is true that the popular reaction to the comet doubtless encouraged the new name of the month to catch on (so Weinstock 157), we know that Quintilis was already being called July in an edict that announced the *ludi Apollinares* (6-13 July), which preceded Octavian's games by several weeks (*Att.* 16.1.1, 4.1).

autobiography,[6] chose to refer to the games as the *ludi Veneris Genetricis*, a name that is regarded by modern scholars as already obsolete by 44 B.C. In fact, *all* sources that link the comet with one of the standing festivals in the Roman calendar report that it appeared when Octavian was celebrating the *ludi Veneris Genetricis* (Texts 1-5), an annual set of games established by Julius Caesar in 46 when he dedicated his temple to the goddess Venus Genetrix (Venus the Ancestress), from whom the Julii claimed to be descended.[7] The celebration in 46, it seems, took place in September since the imperial *fasti* (calendars) assign the anniversary of the temple's dedication to 26 September.[8] This fact, taken with the link in our sources between the comet and a celebration of the *ludi Veneris Genetricis* two years later, has sometimes led scholars to assign Caesar's comet to September 44, the date that astronomers still take for granted.[9] However, the correct date of the Roman sighting of the comet is almost certainly the latter part of July, the date that all classical scholars accept without question.

The dating of the comet to July, rather than September, results from taking into consideration the clear evidence in our ancient sources that the

---

[6] This is a likely inference. Although the name of the games is not specified in the passage quoted by Pliny from Octavian's autobiography (Text 1), Pliny (Text 1A) introduces the passage by stating that Octavian was celebrating the *ludi Veneris Genetricis* when the comet appeared and that Octavian was a member of the Board (*collegium*) established by Caesar to oversee the festival, facts doubtless gleaned by Pliny from the autobiography.

[7] Coinage of the second cent. B.C. attests the link between Venus and the Julii (Weinstock 17), and early in his political career (ca. 68 B.C.) Caesar had advertised the claim of his family's descent from Venus in a funeral oration for his aunt Julia, the widow of Gaius Marius (Suet. Iul. 6.1). For the planning and execution of the design for the new Julian Forum, of which Caesar's temple to Venus Genetrix was the central feature, see Ulrich 49-80.

[8] Degrassi 514. See our Appendix III for a discussion of the probable date of the festival in 46 B.C.

[9] The *communis opinio* among astronomers owes a great deal to the continuing influence of Pingré's eighteenth-century *Cométographie* (a work described by Brian Marsden in his foreword to Kronk as "still very authoritative" but "much out of date"). Pingré (I 277) adopted Sir Edmund Halley's date of 23 Sept. for the comet of 44 B.C., a date Halley (902) arrived at on the basis of "a fragment of an old Roman calendar extant in Gruter p. 135 [= *CIL* I² 219, the *fasti Pinciani*, which attests Augustus' birthday on 23 Sept. and the foundation of the temple of Venus Genetrix on 26 Sept.]". (We thank Brian Marsden for calling our attention to Pingré's indisputable indebtedness to Halley.) The comet is assigned to September by the following scholars (most opting for 23 Sept. and explicitly citing Pingré): Imhoof-Blumer 186, Gundel 1186 (the catalogue of comets readily accessible in *RE*), Chambers 556, Baldet 16, and Hasegawa 65 (the latest, modern catalogue of naked-eye comets from ancient times to A.D. 1970). Two other recent catalogues (Barrett 95-6 and Yeomans 367) do not indicate the month of the Roman sighting in 44.

festival known as the *ludi Veneris Genetricis* was moved from September to July and renamed the *ludi Victoriae Caesaris* in honor of Julius Caesar not many years after its foundation in 46 B.C. Certainly this change had taken place by the early imperial age when the *ludi Victoriae Caesaris* occupied 20-30 July[10] and the *ludi Veneris Genetricis* were no longer being celebrated. Therefore, the correct dating of Caesar's comet by month hinges upon whether the shift of the festival from September to July had taken place in time to affect the celebration in 44 B.C. or occurred somewhat later.[11]

Since classical scholars almost universally assume that the *ludi Veneris Genetricis* were transformed into the *ludi Victoriae Caesaris* and moved to July in 45 B.C., it is taken as a foregone conclusion that Octavian's games in 44 were held in July. To be sure, on this last point there can scarcely be any disagreement. All of the evidence, as we shall see, points to the conclusion that Octavian's games were indeed celebrated in July, rather than September, although we hope to show in the course of this study that 44, and not 45 B.C., was the first year in which the games were celebrated in July. Leaving aside for the moment, however, the specific arguments in favor of 44 as the date of the change in date, this much seems virtually certain: the festival must have been moved from September to July in 44 B.C. at the latest. This conclusion follows from the logical assumption that since the comet was greeted by many (including the future emperor Augustus) as a sign of Caesar's apotheosis, it no doubt insured that the festival during which it appeared would ever afterwards be celebrated on the anniversary of that heavenly portent. This assumption, taken with the fact that two of our ancient sources (Texts 16, 17) indicate that the games in 44 were somehow connected with the *ludi Victoriae Caesaris*, which we know were celebrated in the Augustan age on 20-30 July, virtually guarantees that the games in 44 were held in July.[12]

---

[10] The date is attested by the imperial *fasti* (Degrassi 485-86), and the name is preserved in two of the five calendars that record this festival: "*Lud(i) Vict(oriae) Caesar(is)*", in the *Fasti Maffeiani* (ca. 8 B.C. - A.D. 4) and "*Lud(i) Victor(iae) Caes(aris) divi Iul(i) committuntur*", *Fasti Amiternini* (post ca. A.D. 20).

[11] Unfortunately the reference to Octavian's games in the letters exchanged by Cicero and Gaius Matius not long after the games had been held (Texts 19 and 16 respectively) cannot be used to establish the month of the celebration because the letters could have been written as early as the end of August or as late as mid October (see Shackleton Bailey, intro. n. on *Fam.* 11.27).

[12] We can also establish the probability that the games and comet fell in July, rather than Sept., 44 B.C. by appealing to an argument *ex silentio* based upon Cicero's correspondence. While we shall argue later (pp. 112-16) that Cicero's letters written in July 44 fail to mention the comet and the reaction to it in Rome because at the time Cicero was voyaging to Greece and cut off from news of events in the capital, there is every

By contrast, the evidence is far from compelling for concluding, as most scholars have, that the name of the festival had already been changed from the *ludi Veneris Genetricis* to the *ludi Victoriae Caesaris* by July of 44. Significantly, only the two sources alluded to above connect Octavian's games in 44 with a celebration of the *ludi Victoriae Caesaris*, and the evidence provided by those two texts is potentially ambiguous. One of the two relevant passages is from a letter written by a contemporary witness, Gaius Matius, who was one of the three "agents" (*procuratores*) whose help Octavian enlisted in giving the games.[13] Matius does indeed state that he "superintended the games that young Caesar [Octavian] celebrated in honor of [Julius] Caesar's victory (*Caesaris victoriae*)" (Text 16), but in this context the words "*Caesaris victoriae*" may well be descriptive rather than a formal title for the games.[14] Significantly, the word order is the reverse of the later, official name of the games (*ludi Victoriae Caesaris*). Furthermore, it is not at all certain that Matius was using the expression to refer to the festival as a whole; he may have been referring principally to the funeral games in Caesar's honor (*ludi funebres*) that were of a private nature and were, according to our sources, combined with the public festival in 44.[15] Given the stress that Matius lays on his sense of "duty" (*munus*) to the memory and honor of his deceased friend Julius Caesar, and given Matius' insistence that he was fulfilling a "private obligation" (*privatum officium*), rather than a public function, Matius may have been thinking primarily of the funeral games when he wrote the words "*Caesaris victoriae*".[16] If this was his intent, then Matius implicitly leaves to Octavian the credit for holding the public festival to which the funeral games were attached: the *ludi Veneris Genetricis*, as the festival was apparently called by Octavian himself when he wrote his *Memoirs* two decades later.[17]

The only other text to connect Octavian with a celebration of the *ludi*

---

reason to suppose that the comet would have been attested in the letters if it had appeared in Sept. By Sept. Cicero was back in Rome, and the range of topics treated in the extant correspondence from Sept./Oct. (esp. the attention Cicero pays to how Antony and Octavian were faring in popular opinion) virtually guarantees that the comet would have found a place in the correspondence if it had been a current event in Sept.

13 Attested as *procurator* by Texts 16, 18, and 19.

14 Modern editors (e.g., Shackleton Bailey in his Cambridge ed. and W. S. Watt in his 1982 OCT) apparently so interpret the expression because they do not capitalize "*victoriae*".

15 Text 5, cf. Texts 6-8. For discussion of the evidence and modern scholarship, see pp. 48-50.

16 Weinstock 369 calls attention to the terms *munus* and *procuratores*, pointing out that they are usually applied to gladiatorial contests; therefore, Matius' role may have been connected chiefly with the funeral games for Caesar.

17 Text 1: see p. 3.6 above.

Victoriae Caesaris in 44 is one of two relevant passages in Suetonius (Text 17). Without mentioning the comet, Suetonius relates that Octavian held the *ludi Victoriae Caesaris* when the board charged with producing the festival lacked the courage to do so. Suetonius' account bears some resemblance to Dio's (Text 3), yet according to Dio, Octavian assumed the board's responsibility for giving "the festival that had been introduced at the completion of the temple to Venus." Dio, therefore, must be referring to the *ludi Veneris Genetricis* (not the *ludi Victoriae Caesaris*), and Dio furthermore gives a different explanation for the board's failure to act, saying that the board did not sufficiently honor the memory of Caesar. The one other passage in Suetonius that concerns these games (Text 9) gives quite a different account. It states that the comet appeared at games that were "established in Caesar's honor and celebrated for the first time" by Octavian. In the latter passage, there is no indication that Octavian was acting on behalf of a board whose duty it was to give the festival. [18]

This is the evidence, such as it is, that in July 44 Octavian celebrated a public festival known as the *ludi Victoriae Caesaris*. The overwhelming weight of our evidence suggests, on the contrary, that the games in 44 were principally dedicated to Venus Genetrix (not Victoria Caesaris), and this conclusion flies in the face of the prevailing modern view which assumes that the *ludi Veneris Genetricis* were transferred from September to July in 45 and renamed *ludi Victoriae Caesaris* at that time. [19] There are, however, considerable obstacles that stand in the way of accepting the *communis opinio*.

First, there were no strong incentives in 45 for altering the name and date of the festival that had been established in the previous year in honor of Venus Genetrix. On the contrary, as we shall demonstrate, there were sound reasons for celebrating the *ludi Veneris Genetricis* on its first anniversary in September 45, whereas in 44 it is far easier to identify circumstances that may have led Octavian to advance the date of the festival from September to July.

Second, the sole piece of evidence that is assumed by scholars to attest a celebration of "Victory games" in July 45 (i.e., the *ludi Victoriae Caesaris*) does not necessarily justify that conclusion. The relevant text is a

---

[18] Ever since the sixteenth century, it has been customary to remove the conflict between the two passages in Suetonius by emending Text 9. We shall argue below (pp. 52-54) in defense of the paradosis.

[19] Among those who adopt this view are: Mommsen in *CIL* I^2 322, Degrassi 486, Gelzer 308, Wissowa 456-57, Habel 629, Dar.-Sag. III.2 1378, Weinstock 91, and Rawson 475 (implicitly).

passage from a letter of Cicero to Atticus (*Att.* 13.44) that belongs to a jumbled part of the corpus. Modern scholars have concluded that the letter was written ca. 28 July 45 because of the presumed reference to the *ludi Victoriae Caesaris*, but the adoption of that date in late July makes it necessary to postulate a series of unexpected and unaccountable delays in at least two separate chains of events alluded to in the letter. It is also necessary to shift from July to August, where they fit less satisfactorily, a whole group of letters clearly written after no. 44. None of these difficulties is encountered if letter no. 44 is interpreted as having nothing to do with the *ludi Victoriae Caesaris* and is assigned instead to an earlier date. In that case, there is not one shred of evidence that Victory games were celebrated in July 45.

Third and lastly, if, as scholars generally believe, the name of the games was altered from *ludi Veneris Genetricis* to *ludi Victoriae Caesaris* already in 45, it is certainly odd, to say the least, that not a single one of our sources states that the comet of 44 appeared during the *ludi Victoriae Caesaris*. Instead, all of them link the comet to the *ludi Veneris Genetricis*. Now, given the cumbersome form of Roman dates, the names of recurring annual festivals must have provided a highly convenient way of giving the date of an historical event if it happened to occur during one of those festivals.[20] One would expect, therefore, that by Seneca's day in the first century A.D. (Text 2)—and surely by the time of Dio in the third century (Text 3) and Obsequens in the fourth (Text 4)—the new name for the festival would have ousted the old one, especially if, as modern scholars believe, the games had already been renamed *ludi Victoriae Caesaris* in the year before the comet appeared. By the early imperial period, as we can tell from the *fasti*, there was no longer a festival known as the *ludi Veneris Genetricis*, but everyone would know when the *ludi Victoriae Caesaris* were celebrated. What is the point, then, in resorting to an antiquarian name for a well-known festival if that festival could legitimately be called already in 45 (hence also in 44 B.C.) by the name that was familiar to imperial readers?[21] The retention of the old name in our sources cannot be attributed to religious conservatism because according to the *communis opinio*, Venus Genetrix had already been forced to step aside in 45 B.C. in favor of Victoria Caesaris.

Modern scholars have sometimes dimly perceived this anomaly, but

---

[20] We owe this observation to Michael Alexander, our colleague in History.

[21] To put this in perspective by means of a modern analogy, can we imagine that an American historian writing in the 21st century would adopt the names "Decoration Day" or "Armistice Day", in preference to "Memorial Day" or "Veterans' Day", in giving an account of something that happened on one of those U.S. holidays in the latter half of the 20th century?

they have tended to dismiss it as insignificant by claiming that only one or two sources employ the older name *ludi Veneris Genetricis*.[22] Taking a different tack, Weinstock (157) interpreted the presence of both names in the ancient sources as evidence that some Romans at first refused to accept the renaming of the festival in Caesar's honor. Neither of these explanations is satisfactory. On the one hand, not just a few, but *all* sources that link the comet to a public festival, state that it appeared during the *ludi Veneris Genetricis*. On the other hand, as we have seen (p. 3.6), Octavian himself appears to have called his games the *ludi Veneris Genetricis*, and we can hardly suppose that the adopted son of Caesar refused to call his games the *ludi Victoriae Caesaris* out of any reluctance to see the glory of Caesar further enhanced. This would seem to rule out Weinstock's explanation for the coexistence of the two names, and in our view, the near unanimity of the ancient sources in calling Octavian's games in 44 the *ludi Veneris Genetricis* presents a fatal objection to the theory that the festival was given a new name in 45.[23] In fact, as we shall argue, although the date of the *ludi Veneris Genetricis* was undoubtedly changed by Octavian from September to July in 44, the festival is likely to have continued to be called the *ludi Veneris Genetricis* for some years after that date, becoming only later, in the early Augustan age, the *ludi Victoriae Caesaris*.

The Aims of this Study

Part one of this study seeks to offer a new and more convincing interpretation of the evidence found in our sources which call the games in 44 variously (1) *ludi Veneris Genetricis* (Texts 1-4, 14-15), or (2) *ludi funebres* either on their own (Texts 6-8) or in combination with the *ludi Veneris Genetricis* (Text 5), or (3) *ludi Victoriae Caesaris* (Texts 16-17). To anticipate some of our conclusions, we shall argue that the games established by Caesar in September 46, the *ludi Veneris Genetricis*, were held in the following year also in September and that no change in name was introduced at that time. One

---

[22] For instance, Scott 257 n. 1 refers to "the apparently mistaken remarks of Pliny (*NH* 2.93, = Text 1A) that the comet appeared during games celebrated in honor of Venus Genetrix."

[23] Of course, it might be objected that the agreement of the majority of our sources in linking the comet to the *ludi Veneris Genetricis* (Texts 1-4, cf. 5) simply reflects the dependence of later sources on the official Augustan *interpretatio* of the comet that was published in his *Memoirs* (see pp. 63-65), but if this is so, it merely reinforces our conclusion that Augustus chose to call his games in 44 *ludi Veneris Genetricis*—surely strong evidence that the games had not already been converted into the *ludi Victoriae Caesaris* in 45 as modern scholars insist.

year later, in 44, the date, but not the name, was changed. This will explain why all sources that give the name of the festival during which the comet appeared are unanimous in stating that it was seen during the *ludi Veneris Genetricis*. It is furthermore our belief that in 44 Octavian combined with the *ludi Veneris Genetricis* funeral games (*ludi funebres*) in honor of Julius Caesar. This is specifically stated by one source (Text 5), and *ludi funebres* are attested in three other passages (Texts 6-8). These games in Caesar's honor emphasized his remarkable achievements as a military commander, and we shall argue that Matius (Text 16) referred in particular to the *ludi funebres* when he used a descriptive phrase (*Caesaris victoriae*) that later became the official name of the festival in the early Augustan age.

Octavian appears to have claimed the right to hold the *ludi Veneris Genetricis* "*pro collegio*" (i.e., "on behalf of the board" charged with sponsoring the festival annually, Text 4) both because he was a member of the board (Text 1A) and "on the grounds that the festival was his concern because of his family" (ὡς καὶ προσήκουσαν διὰ τὸ γένος, Text 3)—i.e., his adoption by Julius Caesar caused him to have a special obligation to Venus Genetrix, the ancestress of the Julian gens. By attaching *ludi funebres* for Caesar to the games in honor of Venus, Octavian enhanced the stature of Caesar, his adoptive father, and he could appeal to family precedent to justify his action. When Julius Caesar founded the festival to Venus Genetrix in 46, as part of the first celebration Caesar had included *ludi funebres* in honor of his late daughter Julia (d. 54 B.C.). The combined celebration in 44 drew attention to the link between Venus Genetrix and her descendant, the late dictator, whose honors were allegedly being slighted by some of Caesar's followers after his death, notably by Mark Antony (Texts 14, 15).

We shall argue that the date for this double celebration was determined in part by Octavian's desire to counter the attempt of the tyrannicide Marcus Brutus to court popular favor by giving lavish *ludi Apollinares* (6-13 July) in his capacity as urban praetor (*praetor urbanus*). July also recommended itself to Octavian, no doubt, because it was the month of Caesar's birth, and its name had recently been changed from "Quintilis" to "Iulius" in Caesar's honor. Two separate sets of games in 44, one in July for Caesar and one in September for Venus, were out of the question because Octavian did not have the necessary financial resources. Mark Antony was doing everything in his power to prevent the young Octavian from gaining access to his inheritance from Caesar.[24]

---

[24] Caesar's widow Calpurnia had entrusted Caesar's property to Antony for safe-keeping soon after the Ides of March (Plut. *Ant.* 15.1), and Antony refused to turn over to Octavian the cash that he needed to pay the legacies left to the Roman people under

It is important to investigate the topics outlined above for several reasons. First, we should want to know whether Julius Caesar (the *communis opinio*) or Octavian (our view) deserves the credit for establishing in July the important festival that was known in the imperial age as the *ludi Victoriae Caesaris*. The reasons for modifying the earlier festival in honor of Venus Genetrix and shifting it from September to July will clearly differ, and differ greatly, depending upon whether the changes were made before or after the murder of Caesar. Second, if the games were held in July for the first time in 44 B.C., it means not only that Augustus played a much greater role than has been suspected in shaping the festival and furthering the glory of Caesar, but also that the circumstances of the comet's observation were significantly different from what scholars have assumed. The impression that the sudden appearance of the daylight comet made on the popular imagination was bound to be startling, no matter what its context, but the phenomenon must have seemed all the more portentous if it happened to coincide with a celebration in honor of Venus *and* Caesar that had never before been held at that time of year.

The prevailing view of the two sets of games, in contrast with ours, may be represented schematically as follows:

Standard View

46 B.C.	45 B.C.	44 B.C.
	July 20-30	July 20-30
	ludi Victoriae Caesaris	*ludi Victoriae Caesaris* + *ludi funebres* for Caesar[25]
Sept. 26 ± x days[26] *ludi Veneris Genetricis* + *ludi funebres* for Julia		

View of this Study

46 B.C.	45 B.C.	44 B.C.
		July 20-(30?)[27] *ludi Veneris Genetricis* +*ludi funebres* for Caesar (the latter styled *Caesaris victoriae* by Text 16)
Sept. 26 + x days *ludi Veneris Genetricis* +*ludi funebres* for Julia	Sept. 26 + x days *ludi Veneris Genetricis*	

Caesar's will (Plut. *Ant.* 16.1-2; Appian 3.17.62-64, 20.73-75; Dio 45.5.3-4). Octavian, therefore, was forced to raise cash by selling his own property (Appian 3.21.77-23.89).

[25] Accepted by most, but not all, modern scholars as an element of the games in 44 (see p. 48 and p. 48.25).

[26] Scholars variously treat the games as leading up to, or following the dedication of the temple on 26 Sept., represented here as plus or minus "x" number of days. Since we believe that in both 46 and 45 the festival began on 26 Sept. and lasted into the first part of October, we show this as a plus "x days" under our reconstruction. See Appendix II for a discussion of the date of the temple's dedication, and see Appendix III for the probable date of the games in 46.

[27] The games may not have occupied all of these days in 44: see below, pp. 54-55.

The precise name and past history of the games celebrated by Octavian in 44 are not, however, the only topics that call for discussion. We should also want to know why the comet that appeared during those games was interpreted as a sign of Caesar's apotheosis. Ordinarily comets were regarded in antiquity as extremely baleful signs, and yet the comet of 44 was celebrated at the time, and later by the Augustan poets (Texts 20-24), as a positive omen. It even appears to have contributed to some of Virgil's thinking in the *Fourth Eclogue* where he describes the return of a "Golden Age." Furthermore, this comet may, in fact, have been visible not many months after the Ides of March, much earlier in the year than is indicated by our Greco-Roman sources. Records from China reveal that a comet was seen towards the latter half of May, or first half of June, but there is no direct mention of this sighting in our western sources despite the survival of quite a few contemporary documents. The Roman sighting in July is, of course, well attested in these western sources, but it has left no trace in the extant Chinese sources, nor is Caesar's comet mentioned even in passing in the extant letters of Cicero from July and August 44 B.C. This silence of our sources, both western and East Asian, is troubling and calls for investigation. Finally, there is a highly enigmatic statement attributed to Octavian that has hitherto defied explanation. We are told that although *publicly* Octavian welcomed the comet as a sign of Caesar's apotheosis, *privately* he adopted the view that the comet was a sign meant for him personally: "that it [the comet] had come into being for him and that he was coming into being in it" (Text 1A).

Part two of this study addresses these and related questions. We begin by calling attention to the very significant fact that the comet of 44 B.C. must have seemed to contemporary observers even more portentous than scholars have imagined if we are correct in arguing that the games to Venus had never before been held in July, and if July 44 was the first occasion on which the games were expanded to include honors for Caesar by the addition of *ludi funebres*. By contrast, according to the prevailing view, the comet simply happened to coincide with a festival that had already been celebrated once before (in 45 B.C.) in honor of Julius Caesar. If that is what happened, Octavian and his supporters could, to be sure, adopt the interpretation that the celestial omen was intended to herald the apotheosis of Caesar, but how much more effectively could the coincidence of the games and comet have been exploited if the bright light in the sky seemed to appear miraculously when both the date of the festival and its nature were modified to confer honor on the memory of Caesar, in the month of his birth, a month whose name had only recently been changed to "Iulius"?

The historical reality of the comet can scarcely be doubted, although we are bound to admit that it is a striking coincidence that a comet should appear so soon after Caesar's death, and not only that, but in the month of his birth, during the course of his funeral games. Although most of our western accounts of the comet are based to one degree or another on the official Augustan version of the comet's particulars (see p. 63), the chance survival in several of our sources (Texts 3 and 6) of what appears to an anti-Augustan interpretation of the comet provides evidence that the comet was not just an invention of Octavian's (the future emperor Augustus). Also, it is difficult to imagine why Octavian would go out of his way to invent a comet if none had existed in 44. He would only be borrowing trouble if he alone insisted that such a potentially ill-omened object as comet had appeared during the course of his games for Venus and Caesar when in fact no comet had been seen.

The reality of the comet is also vouched for by our wholly independent Chinese sources which attest a sighting in May-June of 44 B.C. Those sources clearly cannot be suspected of having fallen under the spell of the portentous Ides of March. Although we cannot test the reliability of the notice for 44 B.C., comparable records in the Chinese sources prove to be remarkably reliable where they can be checked, as they can, for instance, in the reports of solar eclipses and the comet that we know today as Halley's Comet. Of course we cannot be certain that the comet seen by the Chinese in the late spring and the comet seen by the Romans in late July were the same comet, but we can show that statistically it is probable that the sightings from China and Rome concern a single comet. If this is true, if there was a single comet in 44, then our Far-Eastern and Greco-Roman sources permit us to determine the likely position of Caesar's comet ca. 30 May and ca. 23 July. By taking into consideration these two positions, as well as certain other conclusions that can be drawn about the probable characteristics of the comet's orbit, it is possible to arrive at a relatively narrow range of orbital parameters that fit our evidence.

These orbital parameters permit us to draw a number of interesting conclusions about Comet Caesar. For instance, we can demonstrate that the comet should have been visible to the naked eye during the early morning hours in June and early July. (There may be just a dim recollection of these sightings in two of our sources [Texts 11 and 12].) It is possible to estimate the comet's brightness during the period between the Chinese sighting in late May and the Roman sighting in late July. This additional information may in turn help explain why the object could have been confused with a star since it appeared fairly continuously over the course of some months but with little or no evidence of a tail. We can furthermore calculate the probable times of rising and setting for the different latitudes of the capitals of Italy and China, and we can demonstrate that the sudden brilliance of the comet in late July must have been unexpected and could not have been foreseen by contemporary observers.

In chapter 5 we try to explain why the sighting in May is attested only in our East Asian sources and why the sighting in July is attested only in our Greco-Roman sources. In discussing the earlier of the two sightings, we draw attention to the unusual atmospheric conditions that are attested in Italy for much of 44 B.C.: the appearance of triple suns, sometimes with a distinctive aureole, and reports of a wholly occluded Sun, which allegedly caused a lengthy period of gloom and even a failure of the crops. These conditions, we shall argue, were most likely caused by an eruption of Mt. Etna that is said to have occurred in the spring of 44. An eruption in 44 B.C. (whether of Etna or of some other volcano) is corroborated by deposits of sulfur dioxide in the ice cores of Greenland, by attenuated tree rings in North America which point to a climate-altering volcanic eruption, and by reports from China of unusual solar and meteorological phenomena. The occluded atmosphere may in part explain the unexpected silence of our western sources for the period May-June when the comet was seen from China. We must also bear in mind that the reports from China are based upon records that were kept by trained observers who were in the service of the emperor. These sky watchers conducted their observations from a tower, with the aid of instruments, and when they caught sight of the comet of 44 B.C. it appears to have been skirting the sunset skyglow, where it would have been extremely difficult to see. This circumstance may also help to explain the silence of our western sources for the period of May-June.

More surprising, at first glance, is the absence of any allusion to Caesar's comet in Cicero's correspondence during the last week in July, when, according to our other sources, the comet was seen from Rome at the 11th hour for seven days and "attracted *everyone's* attention" (*convertit omnium oculos*, Text 4). Significantly, however, Cicero was absent from Rome and wholly cut off from news from the capital from about 17 July until 7 August, the very period during which the comet supposedly caused a major sensation in Rome (20-30 July). Cicero was traveling by ship off the coast of southern Italy when the comet appeared during Octavian's games, and from Cicero's vantage point the comet may not have been a daylight phenomenon because by the time it rose in the NE over the Apennines, darkness may already have fallen. Then too, since Cicero did not observe the comet in the context of Octavian's games and the urban mob's reaction to the heavenly portent, he had no particular reason to comment on it in the extant, contemporary letters. He personally put no stock in the portentous nature of comets and similar phenomena. Finally, since most (though not all) of what our later sources tell us about the comet of 44 is derived in one way or another from the official view of the event that was promoted by Octavian and his supporters, Cicero's silence provides a salutary counterbalance to the notion that the comet was on everyone's lips.

In the final chapter we take a closer look at how the comet was interpreted in antiquity when it first appeared and how it was portrayed in later literature. Two sources (Texts 3 and 6) permit us to see that the comet of 44 clearly inspired at least two rival interpretations in July 44, one of which was decidedly negative and hostile to the view that Octavian wished to promote. We argue that at first Octavian and his supporters denied outright that the celestial phenomenon was in fact a comet, preferring instead to classify it as a new "star." It is as a star, rather than a comet, that the heavenly visitor of 44 is invariably portrayed on Roman coinage during the two and a half decades that followed its appearance, and it is likewise described as a star in Augustan poetry down to ca. 20 B.C. (Texts 20-24). By contrast, Octavian's opponents appear to have insisted that the celestial object was indeed a comet with the usual baleful implications (Texts 3 and 6). Interestingly, this negative interpretation may also be found here and there in the Augustan poets, but it turns up only when the event is disassociated from Caesar's apotheosis and is brought into conjunction with the various portents that supposedly foreshadowed the resumption of civil war after the Ides of March (Texts 31 and 32). Yet a third interpretation may be recovered from the astrological literature on comets; the relevant texts are found in late sources (fourth-sixth centuries A.D.), but they appear to preserve views concerning the nature of comets that evolved in the last two centuries B.C. In those sources, the comet of 44 is specifically classified as a positive omen.

How the positive interpretation of this particular comet evolved is best explained, in our view, by first noticing that when Augustus (Octavian) came to write his *Memoirs* (*De vita sua*) approximately two decades after the comet had appeared, he no longer insisted on calling the comet of 44 a new "star." Instead, in his *Memoirs*, he described it as a *sidus crinitum* (i.e., a "comet"). Most likely the view of the comet that had been adopted by Octavian's enemies in 44 had in the meantime been reinterpreted. Octavian's opponents in 44 appear to have claimed that the comet was to be regarded as a sign that a New Age of the world was dawning. This was said to be the tenth age, which according to Etruscan lore was to be a degenerate period and the final stage in the existence of the world. By contrast, a competing Greek view presented the tenth age as a new Golden Age. Furthermore, since apocalyptic literature quite often insists that such glorious ages must be preceded by periods of great hardship, it is conceivable that when Augustus composed his *Memoirs* ca. 24 B.C., he welcomed the notion that comet of 44 did indeed signal the advent of a new age—not, of course, the one foretold by his enemies in 44 but one that had in the meantime been reinterpreted in a positive light under the influence of the Greek tradition concerning renewal in the tenth age. The time of suffering could readily be identified with the Civil Wars (49-31 B.C.). That turbulent period had been brought to an end by Augustus' victory at Actium in 31, and the succeeding *pax Augusta* could be regarded as the new age foretold by the comet.

Finally, we shall argue that astrology is also likely to have played a role in leading Octavian to adopt his private view that the comet was a sign of his rebirth (Text 1A). Suetonius informs us that Octavian's horoscope had been cast in the spring of 44 before he returned to Italy. Suetonius also states that Octavian regarded Capricorn as the "sign under which he had been born" and that he advertised this fact on silver coins—the latter claim amply confirmed by surviving specimens of Augustan coinage. We shall show that Capricorn was in the ascendant during the hour when the comet rose in July 44. In our view, this circumstance provides a new and promising explanation of Octavian's enigmatic remark which interprets the comet as a sign closely linked with his own birth.

PART I

THE FOUNDATION OF THE *LUDI VICTORIAE CAESARIS*

II

THE PREVAILING VIEW OF THE

CELEBRATION IN 45 B.C.

&

THE EVIDENCE OF *AD ATTICUM* XIII 44

A s indicated in the previous chapter, the festival that Caesar founded in 46 B.C. in honor of Venus Genetrix (the *ludi Veneris Genetricis*) had undergone two major changes by the time of the Augustan age: the games had been moved from September to July, and they had been renamed the *ludi Victoriae Caesaris*. The *ludi Victoriae Caesaris* are firmly attested by the early imperial *fasti* on 20-30 July (Degrassi 485-86), while the sole reminder of the *ludi Veneris Genetricis* is the notice in the Augustan *fasti* that 26 September was the day on which the temple to Venus Genetrix had been dedicated, which we know was the occasion of the inauguration of the annual festival in 46 B.C. (Degrassi 514). Furthermore, although the bulk of our evidence describes Octavian's games in 44 B.C. as the *ludi Veneris Genetricis* (Texts 1-4, 14, 15), two sources (Texts 16 and 17) point to the conclusion that those games were a celebration of the *ludi Victoriae Caesaris*. This identification of the games in 44 with the festival that is known to have been celebrated in July in the Augustan age virtually guarantees that Octavian's games must have been held in July, rather than September. In other words, we can envisage Caesar's new festival to Venus Genetrix being moved from September to July in either 45 or 44, but after the comet appeared at the celebration in 44, the date of the festival must have been fixed once and for all to coincide with the anniversary of the heavenly portent that had been interpreted as a sign of Caesar's apotheosis. In support of this conclusion, all other evidence points to July as the month of Octavian's games; no suitable context can be found for those games in September of 44.

Although the festival could have been moved from September to July conceivably in either 45 or 44, most scholars have adopted the view that the festival received its new name and date in 45.[1] Those who hold this opinion have generally accepted the explanation offered by Mommsen to account for why it would have been appropriate to introduce the changes in 45. Recently, however, Stefan Weinstock has drawn attention to certain glaring weaknesses in Mommsen's explanation and has offered a rival explanation of his own. In fact, the whole question of whether the festival was moved to July in 45 or 44 has assumed greater importance thanks to this imaginative new thesis of Weinstock, according to which the reform in 45 is to be interpreted as a deliberate plan on Caesar's part to lay claim to July (the month of his birth) as his own special month. Weinstock credits Caesar with redirecting the festival from Venus Genetrix to a new personal goddess of Victory (Victoria Caesaris) in order to compete with the god Apollo and the *ludi Apollinares* on 6-13 July. This would be a significant act on Caesar's part to prepare the way for having the month Quintilis renamed "Iulius" in his honor in 44, an honor that clearly set Caesar apart as more than human and contributed to his elevation to "Divus Iulius".

Of course, if it can be shown that the festival in honor of Venus Genetrix was most likely celebrated in September 45 and was first moved to July in 44, then Weinstock's theory falls to the ground. That will be our thesis in the next chapter. Also, if the change in date was introduced in 44, rather than in 45, then Octavian Caesar (Augustus), not Julius Caesar, deserves the credit for causing the *ludi Victoriae Caesaris* to be held in July in imperial times. This would mean that the future emperor Augustus played a much more significant role than has been suspected in shaping the festival that was designed to further the glory of Julius Caesar. And finally, if the festival was first expanded in 44 to include honors for Caesar and moved to July in that year, imagine how much more striking the appearance of a comet must have seemed to contemporary observers. The heavenly omen could readily have been interpreted by many as giving its blessing to the new date of the games and to the extension of that festival to include honors for Caesar in the form of funeral games, to which Matius (Text 16) may allude with the words *victoriae Caesaris* (see p. 5).

Since, however, it is the *communis opinio* that the *ludi Veneris Genetricis* were moved and transformed into the *ludi Victoriae Caesaris* in July 45, let us begin by examining first Mommsen's theory and then

---

[1] See p. 6.19. Gagé 174-75 and Bellemore 121 are among the few scholars who allow for the possibility that the change occurred in either 45 or 44.

Weinstock's to see how each tried to account for why those changes most probably occurred in 45. We shall find that both explanations are vulnerable to serious objections—objections that disappear if the switch from September to July took place in 44, rather than 45. However, before we can present in the next chapter the arguments in favor of 44 B.C., we must first examine a text that has commonly been interpreted as attesting a celebration of the *ludi Victoriae Caesaris* in July 45. The text is found in a letter that Cicero wrote to his friend Atticus in July 45 (*Att.* 13.44), and the allusion in this text to a statue of the goddess Victory being carried in a procession at circus games furnishes the only evidence that the *ludi Victoriae Caesaris* may have been celebrated in July of that year. Obviously we shall have to show that the letter does not concern a celebration of the *ludi Victoriae Caesaris* in July 45 by offering some other, more satisfactory interpretation of the relevant text before we can go on to argue that Octavian Caesar, rather than Julius Caesar, deserves the credit for moving the *ludi Veneris Genetricis* from September to July in 44 and for modifying the festival so that it developed into the imperial *ludi Victoriae Caesaris*.

Mommsen's Theory[2]

Mommsen thought that he detected in Caesar's reform of the calendar in 46 B.C. the likely explanation for why the games to Venus were moved to a new month and day in 45. Mommsen began by observing that when Caesar celebrated the *ludi Veneris Genetricis* for the first time in 46, he held those games on either 24 or 25 September since the imperial *fasti* vary between those two dates.[3] He then went on to observe that 24 / 25 September 46 will

---

2 Accepted by, among others, Dar.-Sag. III.2 1378, Wissowa 456-57, Habel 629, and Degrassi 486.

3 *CIL* I[2] 322, "*fasti enim variant*." The weight of the evidence, however, favors 26 Sept. (*a. d. VI Kal. Oct.*, so *Arv.*, *Pinc.*, *Praen.*), one source only giving 25 Sept. (*a.d. VII Kal. Oct.*, so *Vall.*): see Degrassi 514. (Mommsen in *CIL* I 402 and I[2] 330 misreports *Pinc.* and *Vall.*, whereas in I[2] 219, 240 Mommsen states the facts correctly, in agreement with Degrassi.) Mommsen may have given these dates as 24 / 25 Sept., rather than 25 / 26 Sept., because he was taking into account that Sept. had 29, not 30 days in 46 (so Shackleton Bailey assumes, intro. n. *Fam.* 12.18). It appears more likely, however, that Mommsen misstated the dates "*mero lapsu*", to use Degrassi's phrase (514), since elsewhere in discussing the games in 44 Mommsen states that the anniversary of the temple's dedication fell on 25 Sept., citing *CIL* I 402 where *VII* and *VI Kal. Oct.* are reported as 25 and 14 [*sic*] Sept. respectively (*Gesamm. Schrift.* IV 181, with a correction of 14 to 26 Sept. in the footnote). The error is one that continually gets passed on, thanks to the *auctoritas* of Mommsen (cf. p. 31.21 for another example): Dar.-Sag. III.2 1378, Habel 629.

have corresponded to 23 / 24 July in the Julian calendar,[4] and he speculated that the board appointed by Caesar to oversee the annual celebration of the new festival may have taken the Julian date into consideration when it swore its oath. In other words, the board may have vowed to hold the games in future years on the day in the new Julian calendar that was known to correspond to the day of the festival's inauguration in 46. Finally, Mommsen felt that the change in name (from *Veneris Genetricis* to *Victoriae Caesaris*) was to be explained by assuming that Venus (particularly under the cult title "Victrix"[5]) and Victory could easily be interchanged.[6]

Leaving aside for the moment this last controversial assumption, Mommsen's theory regarding the part played by Caesar's reform of the calendar is open to at least three major objections, the first of which Mommsen himself anticipated. Mommsen had to admit that no other Roman festival is known to have changed its date with the introduction of the Julian calendar, but he argued that the games instituted in 46 could have been treated differently precisely because they were held for the first time in a year that was designed to prepare the way for the new calendar.[7] So, he asserted, it would have been logical for the *collegium* charged with holding the games

---

[4] In 46 B.C., *a. d. VII* and *VI Kal. Oct.* of the civil calendar will have corresponded to 23/24 July (Julian), assuming that the two intercalary months totalling 67 days that were added between Nov. and Dec. 46 caused the Kalends (1st) of January 45 (civil) to fall on 1 January (Julian). In that case, the calendar in 46 will have been 63 days ahead of the Sun. I.e., meteorologically it was summer, which explains why Octavian succumbed to what must have been a heat- or sunstroke, to judge from the severity of his affliction, when he presided over the Greek theatrical performances at the *ludi Veneris Genetricis* in 46 (Nic. Dam. 9.19). The remaining 4 days of the 67 will have been equivalent to the days that were added to three of the autumn months in the Julian calendar (1 day each to Sept. and Nov.; 2 days to Dec.).

25 Sept.–29 Dec. pre-Julian: 5 days in Sept., 31 Oct., 29 Nov., 29 Dec., 67 intercal. = 161 days
24 July–31 Dec. Julian: 8 days in July, 31 Aug., 30 Sept., 31 Oct., 30 Nov., 31 Dec. = 161 days

By contrast, Drumann-Groebe 819-20 (following Holzapfel) calculate that in 46 the calendar was only 62 days ahead of the Sun and so show *a. d. VII* and *VI Kal. Oct.* (civil) = 24 / 25 July (Julian) and place the Kalends of January 45 (civil) on 2 January (Julian). This is also the view of Brind'Amour 123, 346. It makes no difference to our argument which calculation is adopted.

[5] Before the Battle of Pharsalus, it was to Venus Victrix that Caesar had vowed a temple (Appian 2.68.281), and yet in that passage and one other (2.102.424) Appian appears to link Venus "the bringer of victory" with Venus Genetrix. This has encouraged speculation that the two were identical ("*eadem est Venus victrix et Iuliorum genetrix*," Mommsen, *CIL* I^2 323).

[6] Degrassi 486 adopts this view: so too Koch in *RE* 864-65 but not in *Hermes* (1955), 44-45.

[7] Known as "the final year of confusion" (*annus confusionis ultimus*, Macrob. *Sat.* 1.14.3).

annually to take its vow in keeping with the future calendar.

The second and more serious objection is this: Although Mommsen's theory explains how an event that took place in 46 on 25 September (in the pre-Julian, civil calendar) could conceivably have been viewed as having its one-year anniversary on 24 July 45 (in the reformed, Julian calendar), it does not explain why that festival came to occupy 20-30 July in later times. In our view (see Appendix III), the day of the dedication of the temple to Venus Genetrix in 46 provided a link between Caesar's four triumphs and the games that followed the temple's dedication. That day, therefore, should have been treated in future years as the first day of the festival. Yet according to Mommsen's hypothesis, the day in the new Julian calendar that corresponded to this anniversary fell not at the beginning of the games (20 July) but on the fourth or fifth day of the festival (either 23 or 24 July).[8]

The third and final difficulty concerns a related issue and is no less troubling. If the games were moved to 24 July in 45 B.C. because the day on which they had been held for the first time in 46 corresponded to that date in the new, Julian calendar, why was the anniversary of the temple not moved as well? Since the games had originally been celebrated to mark the dedication of the temple to Venus Genetrix, one would have expected the games *and* temple to continue to share a common date and not be split between July and September as they are in the imperial *fasti*.[9]

Weinstock's Theory[10]

Weinstock (156) came to reject Mommsen's explanation for why the games were moved to July chiefly because he was troubled by the first of the three objections discussed above (viz., it presumes a unique consequence of Caesar's reform of the calendar in 46 B.C. since no other Roman festival is known to have been moved to a new date in the first year of the reformed calendar). Weinstock also found unsatisfactory Mommsen's explanation for why the festival received a new name. Instead, Weinstock (91) believed that

---

[8] We cannot very easily resolve this difficulty by simply assuming that the games were extended on either side of 24 July. Ordinarily the principal day of *ludi publici* (i.e., the day especially dedicated to the divinity) came either at the beginning of the festival (so the *Megalenses* and *Florales*), or at its end (so the *Ceriales*, *Apollinares*, *Victoriae Sullanae*). Apart from the much older *ludi Romani* and *Plebeii*, the games never fell on either side of the principal day (Wissowa 455).

[9] Wissowa 457 simply accepts this glaring anomaly and remarks that unlike other *ludi*, the *ludi Victoriae Caesaris* lacked a recognizable "Haupttag" because they became separated from the anniversary of the temple with which they had been associated.

[10] Accepted explicitly by Beaujeu 195 and n. 2.

Caesar caused the *ludi Veneris Genetricis* to be renamed *ludi Caesaris Victoriae* and moved to July in 45 "in order to have that long and splendid festival which was connected with this person in the month which was to bear his name."[11]

The beauty of this theory is that it removes the difficulty of having to explain why the games and the anniversary of the temple of Venus Genetrix no longer shared a common date in the new, Julian calendar (our objection number three above to Mommsen's theory). It also circumvents the second of the three objections that can be raised against Mommsen's reconstruction because, if we follow Weinstock, the date of the games (20-30 July) does not have to be reconciled with a presumed correspondence between 25 (or 26) September and 24 (or 25) July in respectively the pre-Julian, civil calendar of 46 B.C. and the reformed calendar of 45. Both of these difficulties disappear since according to Weinstock's thesis, the foundation date of the temple and its anniversary are not to be viewed as relevant factors in explaining why the games were moved to July and assigned to a specific date. Furthermore, since Weinstock believed that the focus of the festival shifted from Venus to Caesar when the games were moved in 45, he was no longer at a loss to explain why the name of the festival was also changed. Although he had at one time accepted the view that Venus Genetrix, in her guise as Venus Victrix, shared an affinity, or even identity, with the goddess Victory, and so the assimilation of the two goddesses might explain how the games to Venus later became games to "Victoria Caesaris",[12] he later came rightly to reject this hypothesis as untenable. The two goddesses were quite distinct.

While we agree with Weinstock that the festival acquired its new name when the games were converted into a celebration honoring Julius Caesar rather than Venus, we do not believe that this change occurred in 45, during Caesar's lifetime. Nor, in our view, was the new name officially introduced in 44, although it is admittedly foreshadowed in Matius' letter (Text 16). To make his case, Weinstock (152) was forced to assume that the new name for the month Quintilis had already been planned in 45, or even in 46, so that it could be a factor in leading Caesar to hold the games in July 45. In support of this assumption Weinstock cites Appian (2.106.443), but in that passage Appian merely gathers together a list of various extravagant honors that were decreed to Caesar, sometime after his victory in the Spanish campaign. Appian does include in this list the proposal to call Quintilis by the

---

[11] Weinstock 156 n. 5 credits Gagé 175 with being the first to see in the *mensis Iulius* a likely explanation for why the games were moved from September to July.

[12] *RE* 8A.2 (1958), 2514 and, *HTR* (1957), 226.

new name Iulius, but we know that the proposal belongs, in fact, to early 44 (Macrob. *Sat.* 1.12.34) and that the new name had by no means ousted the old one as late as July 44 (*Att.* 16.1.1, 4.1).

The really fatal objection to Weinstock's theory, however—or to any theory, for that matter, that postulates a transformation of the *ludi Veneris Genetricis* into the *ludi Victoriae Caesaris* in July of 45—is that it fails to explain why not a single ancient source links the comet of 44 with the *ludi Victoriae Caesaris*. Instead, all sources that attest both the comet and the name of the public festival during which it appeared, call the festival the *ludi Veneris Genetricis* (Texts 1-4, cf. 5). If those games had already been held once before in July 45 under the name *ludi Victoriae Caesaris*, the name by which they were known in imperial times, it seems incredible that Augustus himself in his autobiography, and later sources, would persist in calling the games in 44 by a name that had already gone out of fashion the year before (according to the *communis opinio*) and had long vanished from the imperial calendar.[13]

THE EVIDENCE OF *AD ATTICUM* XIII 44

There are additional reasons for believing that 45 was *not* the year in which the festival was given a new name and date, but before we present those arguments in the next chapter we first ought to examine the sole piece of evidence that has commonly been interpreted as attesting a celebration of the *ludi Victoriae Caesaris* in July 45. This evidence is found in a letter that Cicero wrote to his friend Atticus in the summer of 45 (*Att.* 13.44.1) and comprises an allusion to a procession at circensian games in which a statue of the goddess Victory was carried. The correct interpretation of *Att.* 13.44 is crucial to our study because if the letter does indeed allude to a celebration of the *ludi Victoriae Caesaris* in July 45, as most scholars believe, then there can be no doubt that the *ludi Veneris Genetricis* had already been moved to July and renamed in Caesar's honor in advance of the games sponsored by Octavian in 44. In that case, some theory such as Mommsen's or Weinstock's must be proposed to account for the changes, and more importantly Octavian's role in 44 was perforce limited to perpetuating an already established set of games. On the other hand, if it can be shown (as we believe it can) that Cicero's letter almost certainly does not allude to the *ludi Victoriae Caesaris*, but rather to another festival held earlier in July, then it remains an open question whether the *ludi Veneris Genetricis* were moved

---

13 See p. 7 for a discussion of the significance of this agreement.

to July in 45 or 44 B.C. It also invites us to consider the possibility that Octavian played a far more important part in giving the festival its new date and expanding it to include honors for Caesar than scholars have previously realized.

Att. 13.44 is also of importance in its own right because it has been used by scholars as a focal point around which to organize chronologically a number of other letters in the correspondence of Cicero. Letter no. 44 is found in a particularly jumbled part of the corpus comprising books 12 and 13 of the *Epistulae ad Atticum*. Many of those letters concern matters of a private nature, and give the appearance of having been added to the collection at a later date, when the compiler could no longer be guided by advice from Atticus or Cicero's secretary Tiro in setting the letters in their proper chronological order.[14] Therefore, although clearly some attempt was made in antiquity to group the letters by topic, many of the groupings, as well as individual letters, are not transmitted by the manuscripts in the order in which they were composed by Cicero. In the course of demonstrating that *Att.* 13.44 does not refer to the *ludi Victoriae Caesaris*, we hope to show that the letter was written nearly two weeks earlier than scholars have assumed, and to use that new fixed point of reference to propose revised and more satisfactory dates for ten or more letters written in July and August of 45.

The chronological sequence of the letters to Atticus in books 12 and 13 that is adopted in modern printed editions tends to rest in large part on the work of O. E. Schmidt.[15] Although modern editors do not agree with all of Schmidt's conclusions, today virtually everyone accepts his view that *Att.* 13.44 alludes to a celebration of the *ludi Victoriae Caesaris* in July 45. Schmidt drew this conclusion from a reference in *Att.* 13.44.1 to a *pompa* ("procession at the circensian games") at which the crowd attending the games gave a chilly reception to a statue of the goddess Victory because of her "bad neighbor" (a veiled allusion to Caesar). On the basis of this presumed reference to the *ludi Victoriae Caesaris* (20-30 July), Schmidt (328-30) assigned letter no. 44 to 20 or 21 July and then proceeded to place in a logical sequence the letters that followed it.

In 1937 Lily Ross Taylor pointed out that Schmidt's date for letter no. 44 was at least a week too early. While agreeing with him that the *pompa* mentioned in letter no. 44 took place at the *ludi Victoriae Caesaris*, Taylor

---

[14] See Shackleton Bailey, *Letters to Atticus* I 69-70.

[15] His views presented in *Briefwechsel* have generally been adopted by subsequent editors of the correspondence: e.g., C. F. W. Mueller (Teubner ed., 1898); Purser (OCT 1903); Tyrrell-Purser (London 1915); Winstedt (Loeb ed. 1918).

(230-31) correctly observed that the *pompa* will have occurred on 27 July, the first day of the circensian games,[16] and not, as Schmidt had supposed, on 20 July since stage shows (*ludi scaenici*) occupied 20–26 July.[17] Therefore, if letter no. 44 does indeed refer to the *ludi Victoriae Caesaris*, it could not have been written earlier than 27 July, and in Taylor's view (231) it was probably composed on the 28th. On the basis of that conclusion, Taylor then proposed new dates for two significant groups of letters with reference to which still others are arranged: (1) those describing Marcus Brutus' journey to confer with Caesar in Gaul in the summer of 45,[18] and (2) those written by Cicero during a brief summer interlude at the coastal town of Astura. Modern editors of Cicero's correspondence generally adopt Taylor's revised dating of *Att.* 13.44,[19] and since the dates of nine or more additional letters to Atticus, and one of the *Epistulae ad Familiares*, ultimately rest upon Taylor's conclusions concerning the content and date of *Att.* 13.44, it is all the more important to examine the passage that has been interpreted as an allusion to a celebration of the *ludi Victoriae Caesaris* in July 45. The relevant text reads as follows:

(1) O suavis tuas litteras!—etsi acerba pompa. . . . populum vero praeclarum, quod propter malum vicinum ne Victoriae quidem ploditur! Brutus apud me fuit; cui quidem valde placebat me aliquid ad Caesarem. Adnueram; sed pompa me deterret. (2) Tu tamen ausus es Varroni dare! Exspecto quid iudicet. Quando autem pelleget? De Attica probo. Est quiddam etiam animum levari cum spectatione tum etiam religionis opinione et fama.

("What a fine letter from you, yet the news about the procession at the circus games was unwelcome. . . . Still, the people did well to withhold their applause from the statue of Victory because of her disreputable neighbor [Caesar]! Brutus paid me a visit, and he was very keen that I should write something to Caesar. I consented, but the subsequent news about the procession causes me to change my mind.

---

16 The imperial *fasti* which attest the *ludi Victoriae Caesaris* designate 27-30 July as *ludi in circo* (Degrassi 484-86), and the *pompa* was always held on the first day of the *ludi in circo* (Regner, *RE* Suppl. 7 [1940] 1627).

17 Schmidt's error is occasionally repeated by modern scholars, presumably out of inadvertence rather than disagreement with Taylor's conclusions; e.g., Weinstock 185 n. 11 who assigns *Att.* 13.44 to 20/21 July.

18 This was the occasion on which Brutus patched up the growing rift between himself and Caesar (being rewarded with a praetorship for 44). Brutus was so won over by Caesar that he sent back word that in his view Caesar was prepared to join the *boni* ("honest men", SB: i.e., those having sound, conservative political views), inspiring Cicero's bitter remark that in order to find any *boni* Caesar would first have to hang himself (*Att.* 13.40.1).

19 E. g., Beaujeu, and Shackleton Bailey in both *Letters to Atticus* V (Cambridge 1966) and his Teubner ed. (1987), although earlier in the OCT (1961) SB inadvertently [?] adopted Schmidt's date of 20 or 21 July: *xiii aut xii Kal. Sext.*

Well then, you have ventured to give [the presentation copy of the *Academica*] to Varro after all! I await his judgment. When will he read through it?

I approve of your decision concerning Attica. There is a certain lift to the spirits to be gained from both the display at the games and the popular notion of their religious nature.")

The key features of letter no. 44 (in order of occurrence) that help to establish its probable date are:

(1) The *acerba pompa* and oblique reference to Caesar as Victory's "*malus vicinus.*"

(2) The mention of a recent visit by Brutus (presumably to Cicero's *Tusculanum*, his estate at Tusculum, a town in Latium c. 20 km. SW of Rome). This visit, we can tell, preceded the news of the *acerba pompa* since, after Cicero learned of the *pompa*, he no longer felt inclined to abide by his commitment to Brutus to write to Caesar. This latter detail makes it clear that Caesar must be the *malus vicinus*.

(3) The acknowledgement that the presentation copy of Cicero's *Academica* had been handed over to Varro, to whom Cicero had dedicated the work. This detail establishes ca. 13 July as the *terminus ante quem non* for letter no. 44 because in letter no. 35-36 (written ca. 13 July) Cicero anticipated that Atticus had already presented the book to Varro but had not yet received confirmation that the book had changed hands.

(4) The fact that both the sight of the spectacle and its *religionis opinione et fama* were expected to have a therapeutic effect on the child Attica, who was just recovering from a recent relapse after a long illness.

Taylor, as we have indicated, assigned this letter to 28 July solely on the basis of the first of the four items enumerated above. She identified the *pompa* with the procession at the *ludi Victoriae Caesaris* on 27 July because of the reference to Victory and her "disreputable neighbor" (Caesar), and so the letter could not have been written before that date. On the other hand, items 2-4 in our list might lead one to conclude that the letter was written approximately two weeks earlier. That is, if letter no. 44 was indeed written as late as 28 July, then it is necessary to postulate an unexpected, two-week delay affecting the presentation of Cicero's *Academica* to Varro and Brutus' departure for Gaul. Furthermore it is surprising to find Cicero adopting such a favorable tone towards the games, recommending their therapeutic effect on little Attica, if they were indeed the *ludi Victoriae Caesaris*, a festival that Cicero doubtless found distasteful. Each of these points taken on its own is not enough to arouse suspicion, but their cumulative effect is such that it is bound to cast doubt on the accepted interpretation of the *pompa*. Let us first, therefore, take a look at the chronological difficulties that result from assigning letter no. 44 to 28 July on the basis of the presumed reference to the *pompa* at the *ludi Victoriae Caesaris* on 27 July. Then, let us see if some other, more suitable context can be found for Cicero's remark about the *pompa*.

Presentation of Cicero's ACADEMICA to Varro

The acknowledgement that Varro had recently received the presentation copy of the *Academica* is a welcome detail because, as we noted above, it establishes 13 July as the *terminus ante quem non* for letter no. 44. When Atticus wrote to Cicero on 12 July, he announced his intention to give the presentation copy to Varro "as soon as he arrived" (*simul ac venerit*, *Att.* 13.35-36). Since Varro had paid Cicero a visit at his *Tusculanum* a few days previously (attested in *Att.* 13.33a.1 of ca. 9 July), and there is nothing in the discussion of the planned presentation to suggest that Varro was expected to be away from Rome for any length of time, it is surprising if Atticus did not make the presentation until close to the end of the month. Certainly Cicero himself expected that Atticus had already handed over the copy when he wrote to Atticus (ca. 13 July) "and so the manuscript has now been delivered to Varro, and you no longer have a free hand" (*dati igitur iam sunt nec tibi integrum est*, *Att.* 13.35-36), although Schmidt (329) correctly observed that Cicero's words do not necessarily mean that the presentation (as later reported in ep. 44.2) had already taken place. Cicero indicates as much when he muses that his previous letter of ca. 12 July (*Att.* 13.25.2) may have caused Atticus to change his mind after expressing the intention in his letter of the same day to give the copy to Varro "as soon as he arrived." However, a delay of more than two weeks (from 12 to 28 July) must have intervened, as Taylor (231) confessed, if letter no. 44 acknowledging the delivery was written on 28 July.

Brutus' Journey

Similarly, the reported visit of Brutus to Cicero's *Tusculanum* forces us to revise our estimate of Brutus' probable date of departure for Gaul by a like period of approximately two weeks if letter no. 44 was not written soon after letter no. 35-36 (ca. 13 July) but is assigned to 28 July. Recognizing this fact, Taylor reassigned the group of letters in book 13 that mention Brutus' presence at Caesar's headquarters in Gaul (nos. 38, 39, and 40) to a later date than scholars had previously assumed since time had to be allowed for

Brutus to reach Gaul and for news of his conversations to begin to filter back to Rome. By the date of letter no. 38 Cicero had learned from many persons who were returning to Rome that Brutus was replying eloquently to Cicero's critics at Caesar's headquarters. (Chief among those detractors was Cicero's young nephew Quintus.) According to letter no. 39, Atticus had mentioned the possibility that Brutus would soon be back in Rome after just a brief absence at Caesar's headquarters. Taylor, therefore, assigned letters nos. 38, 39, and 40 to 15, 16, 17 August, a revision of Schmidt's dates (4, 5, 7/8 Aug.) made necessary by the shift of letter no. 44 to 28 July from 20 or 21 July, where Schmidt had wished to place it.[20]

It is indeed surprising, however, to find Brutus still lingering in the vicinity of Rome as late as 28 July, assuming that letter no. 44 was written on that date. Brutus' anticipated departure is first mentioned circa 22 June (*Att*. 13.11.2). In a letter written approximately two weeks later (*Att*. 13.33a.2 of ca. 8/9 July) Cicero withdrew his request for Brutus to be present at a business transaction on 15 July so as not to inconvenience him and hold him up. Just a few days later, Cicero (*Att*. 13.25.2 of ca. 12 July) implies that Brutus was expected to set out towards the middle of the month. As Shackleton Bailey notes in his comment on *Att*. 13.23.1 (ca. 10 July) where Brutus is said to be faced with a *subitum* (implying urgency?) *et longum iter*, this journey must have been postponed until the end of the month if letter no. 44 is assigned to 28 July.

Letters from Astura

These two interruptions in the expected flow of events, each amounting to approximately two weeks, should arouse our suspicion. But that is not all. There is the above mentioned second group of letters that have to be assigned to new dates if *Att*. 13.44 was written on 28 July, as Taylor argued it was. The four letters in this group (three from *Att*. 13: nos. 21, 34, and 47a, together with *Fam*. 6.19) were written from the coastal town Astura (S of modern Anzio, approx. 65 km. SW of Rome) during the last week of either July or August 45. Two of these letters give the days, without specifying the month, on which Cicero traveled to and from Astura: departure from Tusculum for Astura on the 25th (ep. 21.2); planned return from Astura to Rome on the 31st (ep. 47a.1). Schmidt was able to place the Astura letters at the end of July, where they make better sense than they do at the end of August,

---

[20] M. Gelzer, of course, in his article on Marcus Brutus, adopted Schmidt's chronology: *RE* 10.1 (1917), 986.

because he had assigned letter no. 44 to 20 or 21 July on the false assumption that the *pompa* referred to had taken place on 20 July. Of course, if we accept Taylor's date for letter no. 44, that puts Cicero at Tusculum on 28 July, and so the sojourn at Astura from the 25th and 31st of the month must have taken place in August, not July. This conclusion is inevitable, and yet it forces us to suppose that Cicero set out for Astura precisely at a time when we would have expected him to remain in or close to Rome, at Tusculum.

Several considerations make it much easier to envisage Cicero making his journey from Tusculum to Astura at the end of July, rather than at the end of August. To begin with, in late August Cicero had invited his nephew, Quintus junior, to visit him at his *Tusculanum* on 25 August (*Att.* 13.51.2 of 24 August), and yet according to Taylor's chronology 25 August must have been the day of Cicero's departure for Astura. Taylor (236) had to admit that there must have been a "sudden change in Cicero's plans." Next, Cicero was most anxious throughout the summer not to be caught away from Rome when Caesar returned from his Spanish campaign. In July, Cicero was expecting Caesar to return "not before 1 August" (*non ante Kal. Sext.*, *Att.* 13.21a.3 of 30 June or 1 July), and a stay at Astura in late July would not have conflicted with his intention to be in Rome on the day of Caesar's arrival. By mid to late August, however, two conditions had changed. First of all, the date of Caesar's expected arrival was more precisely known: according to *Att.* 13.45.1 (of ca. 11 August) and ep. 46.2 (of ca. 12 August), Caesar was to arrive before the *ludi Romani* (4-18 September).[21] Secondly, by later in the summer Caesar's probable route must have been known since Cicero toyed with the idea of going north to Alsium (approx. 35 km. NW of Rome on the *via Aurelia*) to meet him (*Att.* 13.50.4 of 23 August). In contrast with the certainty expressed in these letters of mid to late August, one of the letters written from Astura (*Fam.* 6.19.2) indicates that Cicero lacked information concerning both Caesar's route and probable date of arrival—quite an understandable state of affairs if the letter was written in late July, but

---

21 Not 5-19 September as Taylor (232) and Shackleton Bailey (on *Att.* 3.45.1), following Mommsen (*CIL* I² 328-29), assert. The question hinges on whether the 5th or the 19th was the sixteenth day that was proposed as an addition to the *ludi Romani* in 44. Since we are told explicitly that this new day was to be a fifth *dies circensis* (*Phil.* 2.110), it must have been added to the end of the festival, on the 19th, when the *ludi in circo* were held (so Wissowa 454, Denniston, 169-70, and Degrassi 507). Shackleton Bailey appears to take this fact into account in his note on *Phil.* 2.110 where he describes the "final stage of the *ludi Romani*" as "lasting from 15 to 18 Sept." This means that (pace Taylor 237) the reference to the Nones (5 Sept.) in one of the letters from Astura (*Att.* 13.34) has nothing to do with the first day of the *ludi Romani*, since those games began not on the Nones but on *prid. Non. Sept.* (4 Sept.), and furthermore the expected date of Caesar's arrival (*ante ludos Romanos*) was in any case clearly sometime before the Nones.

puzzling if it was written at the end of August where it is put by Taylor and modern editors. Then too, in the letters of mid to late August Cicero specifically states that he will soon have to return to Rome, that even Tusculum (approx. 20 km. from Rome) is too far afield, because he wants to be sure to be in Rome to greet Caesar (*Att.* 13.51.2 of ca. 24 August). Such statements are in direct conflict with the plans outlined in the final letter written from Astura (*Att.* 13.47a); in that letter Cicero reveals that he will make just a brief visit to Rome on the 1st (Aug. / Sept.?) to attend a meeting of the Senate and then proceed on that same day to his estate at Tusculum, where he looks forward to an extended stay.

Of course Caesar may have changed his plans and let it be known that he was no longer expecting to arrive "before the Roman games", i.e., before 4 September.[22] In that event, Cicero could snatch a few days at the end of August for a visit to Astura and spend the balance of those days at his *Tusculanum*. Such an assumption is *possible* , but, as we have seen, it is by no means the only change in plans, or postponement of imminent events, that must be imagined in order to assign letter no. 44 to 28 July. That date, in turn, rests solely on the assumption that the *pompa* to which letter no. 44 refers was the one at the *ludi Victoriae Caesaris*, which in imperial times took place on 27 July. Since, however, the balance of the evidence furnished by the letter points to an earlier date of composition, it is worth asking whether the *pompa* could, in fact, belong to an earlier occasion. Specifically, let us explore the possibility that Cicero may have been referring to the *pompa* at the *ludi Apollinares* on 13 July, as opposed to the *pompa* at the *ludi Victoriae Caesaris* (held on 27 July, according to the imperial *fasti*).

The Ludi Apollinares in 45 B.C.

The prevailing interpretation of letter no. 44 is based upon two assumptions which have led to the conclusion that the games in question were the *ludi Victoriae Caesaris*. It is assumed (1) that the cool reception received by the goddess Victory presupposes that the games were being celebrated in Victory's honor; and (2) that this cool reception was caused by the fact that Caesar's ivory statue was being carried in the procession, side by side with the statue of Victory. This circumstance, in turn, is said to account for the oblique reference to Caesar as Victory's "neighbor" (*vicinus*). In our view, however, not only is there another, more probable

---

[22] So Shackleton Bailey assumes in his introductory note on *Att.* 13.34. However, Caesar is attested just outside of Rome on 13 Sept. (Suet. *Iul.* 83.1).

explanation for the reference to Caesar as Victory's "neighbor" (see Appendix IV), but the two main assumptions are based upon false reasoning. Granted that Caesar's ivory statue was do doubt carried in the procession that Atticus described, and the presence of that statue among the statues of the gods may have in part inspired the crowd's chilly reception of the procession, we believe that there is a further, more compelling reason why the crowd's reaction to the games was judged by Cicero and his friend Atticus as a comment on Caesar's popularity. We also believe that the prominence given to the goddess Victory in Cicero's comment does not necessitate the conclusion that the games were "victory games". Instead, we believe that Atticus was most likely describing the reaction of the crowd to the *pompa* at the *ludi Apollinares* on 13 July, a festival in which Caesar played a key role in 45 B.C.

In order to put the prominence given to the goddess Victory in its proper context, it is important to recall that a winged statue of Victory is likely to have been carried at the head of the *pompa* in *all* games.[23] These *pompae* had many features of a triumphal procession following victories in war, a pattern first established when the *ludi Romani* were made an annual festival in September. The *ludi Romani* developed out of *ludi votivi* that were commonly celebrated at the conclusion of a successful season of military campaigns (Degrassi 507), and the form of the *ludi Romani* in turn influenced the other festivals that were added in course of time (Marquardt 508). Triumphal trappings (*vestis triumphalis*, *tunica palmata*, *toga picta*, and crown held by a *servus publicus*) are attested for games under the Empire and hinted at in republican times.[24] Furthermore, the goddess Victory is likely to have played a significant role especially in the celebration of the *ludi Apollinares* in view of their origin. According to Livy (25.12.15), when the *Apollinares* were first celebrated in 212 their aim was to secure victory in the Second Punic War, not relief from pestilence (*victoriae*, *non valetudinis ergo*), and so *Victoria* may well have traditionally enjoyed a prominent place in the *pompa* of these games honoring Apollo.

Therefore, we have grounds for believing not only that the goddess Victory was carried at the head of *all* circus processions (not just at *ludi Victoriae*) but that she may have enjoyed a special connection with the *ludi*

---

[23] Ov. *Am*. 3.2.45 *prima loco fertur passis Victoria pennis* ("first in the procession Victory is carried with wings spread"); cf. the representation of Victoria immediately behind the Magna Mater in a *pompa* portrayed on a sarcophagus, Dar.-Sag. I.2 1193, fig. 1528.

[24] Juv. 10.36-46, cf. 11.193-95, Mart. 8.33.1; praetor's use of *biga* apparently in the *ludi Apollinares*, Pliny *NH* 34.20; for the republic, see Livy 5.41.2.

Apollinares. Moreover, since the entry of the *pompa* into the circus marked the opening of the *ludi in circo*, the statue of Victory at the head of the procession would inevitably be the first to be greeted by the crowd's approval (or disapproval) as signified by the intensity of their applause. This reason alone may account for the prominence given to Victory in Cicero's account: Victory could be described as the recipient of the chilly reception, whether or not the games were being celebrated in Victory's honor, and significantly Cicero adds that the reaction was caused by Victory's *malus vicinus* (viz., Caesar). In other words, the unenthusiastic greeting accorded to the statue of Victory at the head of the procession, which contained the statues of all the other gods and goddesses, could somehow be interpreted as a sign of the crowd's displeasure with Caesar.

This last point need not imply that the games in question were the *ludi Victoriae Caesaris*; an equally, if not more suitable context for the reaction of the crowd to be interpreted by Cicero and Atticus as a barometer of Caesar's popularity can be found earlier in the month of July, at the celebration of the *ludi Apollinares*. The *pompa* on 13 July at the *ludi Apollinares* in 45 B.C. was most likely only the second occasion on which Caesar's ivory statue was carried at the games, a new and exceptional honor that the Senate had granted to Caesar in April in response to the news of his victory at Munda.[25] One earlier occasion is alluded to in another letter.[26] This earlier procession is apparently overlooked by Taylor (230), or else she fails to take into account the role played by the winged statue of Victory in all circensian processions, since she refers to the *pompa* of *Att.* 13.44 as "the occasion on which, it would seem for the first time Caesar's image accompanied by that of Victoria appeared in the circus procession with the images of the gods."

Since Cicero alludes in his letter of May to the previous occasion, it

---

[25] Dio 43.45.2; cf. Appian 2.106.440; *Att.* 14.14.1, ep. 19.3.

[26] *Att.* 13.28.3 of 26 May: *hunc de pompa, Quirini contubernalem* ("this person whose statue is carried in the *pompa*, the fellow-lodger of the god Quirinus"). This occasion is usually said to have been at the *Parilia* on 21 April (so Shackleton Bailey ad loc.; Weinstock 110). Might not a better occasion, however, be found in the month in which the letter was written, on the final day of the *Florales*, 3 May, a *dies circensis* (*Fast. Ven. & Esqu.*: see Degrassi 449)? In 45 the Senate's decree establishing circensian games on the *Parilia* in Caesar's honor was not passed until the news of Caesar's victory arrived on the evening of the preceding day (Dio 43.42.3; cf. Appian 2.106.440). This allows little time for suitable arrangements, unless all was done behind the scenes, well in advance. We cannot, of course, rule out such secret arrangements, especially if we are prepared to assume with Drumann-Groebe (580 n. 3; so also Weinstock 175-76) that Caesar's trusted advisors (esp. Balbus and Oppius) deliberately postponed the public announcement of the victory so that it would fall on the eve of the celebration honoring Rome's founding.

should not surprise us to find Atticus and Cicero taking an interest in the crowd's reaction on the next occasion when Caesar's statue will have been carried in a *pompa*. That next occasion was 13 July, a day having added significance because it was Caesar's birthday. Either in 45, or perhaps more likely early in the following year, the Senate decreed that public sacrifices were to be performed on the day of Caesar's birth. [27] The effect of this decree was to shift some of the attention away from Apollo to Caesar on 13 July, the most important day of the *ludi Apollinares*. This encroachment by Caesar on Apollo's honors apparently stirred protest, and it may have been in 44 that a Sibylline oracle was brought to light which forbade honors to any god except Apollo on 13 July (Dio 47.18.5-6). [28] The matter was eventually settled under the triumvirs by shifting the sacrifices to Caesar from 13 to 12 July, where they were made obligatory for all citizens (Dio loc. cit.). We can identify, therefore, at least two possible reasons why Cicero and Atticus may have interpreted the crowd's reaction to the *pompa* at the *ludi Apollinares* on 13 July as being caused by Caesar (*propter malum vicinum*): (1) Because the *pompa* was being held on a day on which honors were possibly being paid to Caesar at the public altars, and (2) Because Caesar was certainly being honored in the procession itself which included his ivory statue for only the second time. [29]

Finally, unique circumstances in 45 caused the *ludi Apollinares* to have an especially close connection with Caesar in that year, and this connection provides a third reason why Cicero and Atticus may have interpreted the crowd's chilly reception of the procession as reflecting their displeasure with Caesar. Ordinarily the games to Apollo were sponsored by the *praetor urbanus*, but since no praetors were elected for 45 until Caesar returned from Spain in September, the games in 45 were held by one of Caesar's city prefects, at Caesar's expense. [30] In effect, therefore, Caesar was the sponsor

---

[27] Dio 44.4.4. Weinstock 200 assigns this decree to 45, citing Appian 2.106.442; cf. p. 206. Taylor, *Divinity* 67 with greater probability assigns it to 44.

[28] So Taylor, *Divinity* 90 n. 20.

[29] The withholding of applause on this occasion in July is usually said to be the incident referred to in Cicero's *Pro rege Deiotaro* 33 of Nov. 45 (so Halm-Laubmann ad loc. and Gelzer 308 n. 2). Cicero's further description of the incident (§34), however, rather suggests that he may have been alluding to another occasion on which Caesar experienced a cool reception *in person*, perhaps in the autumn of 45 at the *ludi Romani*(?), which Caesar had promised Lamia he would attend (*Att.* 13.45.1).

[30] Dio 43.48.3. Dio (43.48.4) also mentions that a special decree authorized the plebeian aediles to hold the *ludi Megalenses* in April. Only plebeian aediles and tribunes had been elected in 46 before Caesar's departure for Spain (Suet. *Iul.* 76.2), and normally the *Megalenses* were the responsibility of the curule aediles, while the plebeian aediles had charge of the *Ceriales* and possibly the *Florales* (Wissowa 456 n. 2).

in lieu of the *praetor urbanus*, and so the reaction of the crowd to the *ludi Apollinares* could be viewed as a measure of Caesar's popularity. Accordingly, there was good reason for Atticus to send Cicero news of how the crowd greeted the *pompa* at the *ludi Apollinares* in particular. As an example of a comparable report, Cicero drew attention to the popular demonstrations at the *ludi Apollinares* in 44 as proof that Marcus Brutus, the sponsor of the games, enjoyed great favor with the Roman people (*Phil.* 1.36). The circumstances were similar in that both Caesar and Brutus were unable to be present at the games they sponsored (each had a surrogate), and so the reaction of the crowd could be viewed as especially significant in gauging their relative popularity with the Roman people.[31]

Once we realize that there are sound and cogent reasons for believing that the *pompa* mentioned in letter no. 44 may have been the one at the *ludi Apollinares*, we can more easily understand Cicero's comment on how the convalescent Attica might experience a lift in her spirits from attending the show. Indeed his comment provides one more supporting argument for the view that Atticus had written, not about the *ludi Victoriae Caesaris*, but about the *ludi Apollinares*. If the reference to Victory is interpreted as evidence that the *ludi Veneris Genetricis* (founded by Caesar in 46) were moved to July and renamed *ludi Victoriae Caesaris* in 45, then the festival was not only a recent creation but was still in the process of refinement.[32] While Cicero could conceivably have had the new festival in mind in his comment on the pleasure to be derived from "viewing" it,[33] his remark about the *religio* of the occasion better suits the *ludi Apollinares*, dating back to the third century. It is hard to imagine Cicero lauding even the "popular notion of its religious quality" (tr. Shackleton Bailey) if the celebration was such a recent innovation in honor of "Caesar's Victory", clearly distasteful to both Cicero and his correspondent.[34]

---

[31] For the importance of popular demonstrations at the games as a indicator of public opinion, see Cic. *Sest.* 105-106, 115-16, 121. Interestingly several of the demonstrations reported in the letters to Atticus concern specifically the *Apollinares* (Cicero greeted with applause in 54, *Att.* 4.15.6; signs of disfavor shown to Caesar and Pompey in 59, *Att.* 2.19.3; add *Att.* 1.16.11 since the favorable reception of Cicero at games in 61 may have reference to the *Apollinares*). Possibly demonstrations at the *Apollinares* came to be viewed as particularly indicative of the Roman people's sentiment because that festival was held in July, close to the elections when many citizens from the Italian towns and countryside would have been in Rome.

[32] According to Weinstock 91, *Victoria Caesaris* was "created [in 45] to be a personal goddess of Caesar."

[33] The only meaning of *spectatio* recognized by the *OLD* (s.v. 1) in such a context. "Pageantry" (so Shackleton Bailey) is probably best reserved for *spectaculum*.

[34] Schmidt 330, in fact, supposed that Cicero's desire to avoid being in Rome at the

The Date of AD ATTICUM XIII 44

It remains, now, to suggest an appropriate date for letter no. 44 and to give a revised chronology for Cicero's movements through the end of July. Since, according to our view, the *pompa* about which Atticus wrote took place on 13 July, we should expect both his letter and Cicero's reply (letter no. 44) to have been written close to that date. Letter no. 35-36, as we have noted, was written ca. 13 July, prior to the presentation of the *Academica* to Varro, and so provides a *terminus ante quem non* for letter no. 44. Room must be found, however, for one other letter between no. 35-36 and no. 44. This letter, no. 43, is a short note acknowledging a letter that Atticus wrote *ab ludis* ("from the games"), informing Cicero that he could postpone for a day or two his visit to Rome which had been planned for 14 July. It reads as follows:

Ego vero utar prorogatione diei, tuque humanissime fecisti qui me certiorem feceris, atque ita ut eo tempore acciperem litteras quo non exspectarem, tuque ut ab ludis scriberes. Sunt omnino mihi quaedam agenda Romae, sed consequemur biduo post.

("I shall indeed take advantage of the postponement [of some matter connected with Cicero's projected visit to Rome on 14 July]. It was extremely kind of you to let me know and to do so in such a way that I received a letter from you when I was not expecting one, and that on top of this you wrote while attending the games. There are, to be sure, some things that I must do in Rome, but I shall accomplish them two days later than planned.")

Since it is commonly assumed that Atticus' letter *ab ludis* was written on 13 July (to which letter no. 43 replies), there appears to be no room left for Atticus to write a separate letter describing the *pompa* at the *ludi Apollinares* on 13 July. It was, in fact, this very difficulty that led Schmidt to reject Schiche's view that the *pompa* mentioned in letter no. 44 was the one at the *ludi Apollinares* (Schiche 608). To make the connection with the *ludi Apollinares*, Schiche had proposed combining letters nos. 43 and 44 into a single letter so that Atticus' letter *ab ludis* could also be the one in which he described the *pompa*. Schmidt (329), however, cogently argues that letters nos. 43 and 44 cannot so easily be combined. Each must be a reply to separate letters written by Atticus, one *ab ludis* and the other describing the *pompa* and delivery of Cicero's *Academica* to Varro.

The solution to the problem, we believe, lies in perceiving that Atticus' letter *ab ludis* was not necessarily written on 13 July, as is generally assumed (e.g., by Taylor 230). Since *ludi scaenici* ("shows in the theater")

time of the *ludi Victoriae Caesaris* might have caused him to return to Tusculum after making the brief visit to Rome that was planned for 16 July (*Att.* 13.43).

were presented on 6-12 July, Atticus could just as well have written a brief note *ab ludis* on 12 July, and it could have been received by Cicero after he had already written and dispatched his reply (letter no. 35-36 of 13 July) to a letter presumably also written by Atticus on the 12th. (We know that in this period of almost daily correspondence between Rome and Tusculum, Cicero sometimes received two letters written by Atticus on the same day, one in the morning and another in the evening, e.g., *Att.* 13.23.1, and so there is no difficulty in imaging that this happened on 12 July.) Atticus' letter *ab ludis* must have arrived *after* Cicero wrote letter no. 25 (ca. 12 July) stating his intention to go to Rome on the 14th and *before* his projected date of departure on the 14th since the whole purpose of Atticus' letter *ab ludis* was to inform Cicero that he need not come to Rome on the 14th. Furthermore, Cicero specifically states that he received Atticus' letter *ab ludis* at a time "when he was not expecting one" (*quo non exspectarem*)—understandable if it arrived close on the heels of another letter written ca. 12 July to which letter no. 35-36 had already been dispatched as a reply. Room for letter no. 43, therefore, can be found either late on 13 July or early on the 14th (presumably before Brutus' visit alluded to in letter no. 44), the date to which letter no. 43 is traditionally assigned. The brevity of letter no. 43, however, points to the evening of 13 July, after Cicero had already sent any news of substance in letter no. 35-36 earlier on the same day.

Under this scheme, Atticus' letter to which letter no. 44 replies must have been written late in the afternoon, or on the evening, of the 13th at the earliest (if the *pompa* was the one at the *Apollinares* on the 13th) and at the latest a day or two before the revised date of Cicero's projected visit to Rome, 16 July. Most likely it was written closer to the beginning than to the end of this interval since Atticus had two significant pieces of news to pass along right away: (1) the presentation of the *Academica* to Varro and (2) the reaction of the crowd on 13 July. Brutus' visit to Tusculum had best be assigned to the 14th since it must precede the arrival of Atticus' letter describing the *pompa* on the 13th, which caused Cicero to change his mind about heeding Brutus' encouragement to write to Caesar. Room can be found, therefore, for letter no. 44 on the evening of 14 July (or morning of the 15th, the day before Cicero's likely departure for Rome).[35]

---

[35] The request in letter no. 44.3 to be sent a book may, at first glance, appear to be an obstacle to dating the letter to 14 July, just a day or two before Cicero's projected visit to Rome. However, communication by courier between Rome and Tusculum was quite rapid: e.g., the final draft of the *Academica* was being prepared ca. 10 July according to *Att.* 13.23.2 and was in Atticus' hands ca. 11 July according to *Att.* 13.24.1; *Att.* 13.25.3 of ca. 12 July indicates that Cicero had already received Atticus' reply to the previous letter, no. 24, written on the 11th. Therefore, it may have been faster and more convenient to request

Revised Chronology of Mid-July to 1 August

The reinterpretation of letter no. 44 offered here and the new dates proposed for letters nos. 43 and 44 lead to the following reconstruction of events from 12 July through 1 August:

12 July: (**A.M.**) Atticus' FIRST LETTER reports his intention to present Cicero's *Academica* to Varro *simul ac venerit* ("as soon as he arrives"). Atticus has not yet received letter no. 25, written on 12 July.

(**afternoon**) Atticus learns from letter no. 25 of Cicero's intention to visit Rome on 14th[36] and writes a SECOND, unexpected LETTER *ab ludis* (*scaenicis* of *ludi Apollinares*), permitting Cicero to postpone his visit.

13 July: (**during the day**) Cic. writes letter no. 35-36 in response to Atticus' FIRST LETTER of the 12th.

Atticus presents Cicero's *Academica* to Varro; Atticus and his daughter attend the circensian games of the *ludi Apollinares*.

(**evening**) Cic. replies by brief note (13.43), his second of the same day, to Atticus' SECOND LETTER of the 12th, written *ab ludis*.

(**evening**) Atticus writes of *pompa* and presentation of the *Academica* to Varro

14 July: Cic. entertains Brutus at Tusculum and later that day receives Atticus' letter written on the evening of the 13th.

(**evening**)   Cic. (13.44) reports on Brutus' visit and answers Atticus' letter of the 13th concerning the *pompa*.

16-24 July: Cic. visits Rome and returns to Tusculum

25 July: Cic. leaves Tusculum for Astura (*Att.* 13.21.2). If this letter is assigned to August (so Taylor), Cic. leaves unexpectedly on the date of the invitation he had extended to young Quintus. An Aug. date, therefore, forces us to assume with Taylor (236) a "sudden change in Cicero's plans."

26 July: Cic. at Astura, hopes to put off his return to Rome until Nones (*Att.* 13.34)

---

the book by letter, a day or two before Cicero set out for Rome, than to present the request in person.

36 Originally, Cicero had planned to be in Rome for the auction of the Scapulan *horti* on 15 July, but when he decided not to bid on that property, the date of his visit was to be at Atticus' convenience (*Att.* 13.33a.1: see Shackleton Bailey ad loc.).

26-29 July: (**1.**) Wrote *Fam.* 6.19 to Lepta: will remain until he learns more definitely about Caesar's anticipated date of arrival—not before 1 Aug. according to *Att.* 13.21a.3. Cic. has not yet been able to speak to Balbus about Lepta's desired commission because of Balbus' gout (cf. *Att.* 13.4a.1 *Balbus est aeger*, also written at Astura). By contrast, *Att.* 13.46 (securely dated to ca. 12 Aug.) informs us that Lepta, out of eagerness to secure his commission, brought Cic. and Balbus together.

(**2.**) Wrote *Att.* 13.21: the mention (§2) of the need to secure Dolabella's help in reconciling Caesar with Torquatus looks forward to *Att.* 13.45.2 (securely dated to ca. 11 Aug).

30 July: Cic. leaves Astura; will be in Rome on the 31st; will attend Senate on the 1st (Aug.), as requested by Lepidus (*Att.* 13.47a).

1 Aug.: Back at Tusculum (*Att.* 13.47a).

III

THE TRANSFORMATION OF
THE *LUDI VENERIS GENETRICIS*
INTO THE *LUDI VICTORIAE CAESARIS*

The Arguments against Moving the Games in 45 B.C.

I f we were correct in arguing in the previous chapter that *Att.* 13.44 refers
to the *pompa* at the *ludi Apollinares* on 13 July, then there exists not one
shred of evidence that the *ludi Victoriae Caesaris* were held in July 45. In
fact, the earliest attested celebration of those games is now the celebration
by Octavian in July 44, if we may draw this conclusion from the phrase
"*Caesaris victoriae*" in Matius' letter (Text 16) and from the outright
assertion of Suetonius (Text 17). The bulk of our sources, however, (Texts
1-5, 14, 15) would lead us to believe that the games were still being called
ludi Veneris Genetricis in 44. That is apparently how they were described by
Augustus himself in his autobiography (see pp. 3.6, 6.23). Furthermore, as
we have indicated in the previous chapter, we can be reasonably certain that
the festival had been moved from September to July by the year 44 at the
latest, but now, thanks to the revised interpretation of *Att.* 13.44, we are no
longer bound to assume that the *ludi Veneris Genetricis* were transformed
into the *ludi Victoriae Caesaris* and moved from September to July in 45 B.C.
In fact, if we examine the two years in which the change from September to
July could have been made, we shall find that circumstances in 45 B.C.
favored holding the festival once more in September, the month in which it
had been inaugurated in 46, whereas in 44 B.C., conditions were ripe for
holding in July games that were to be the forerunner of the imperial *ludi
Victoriae Caesaris* (20-30 July).

To begin with, in 45 Caesar was away from Rome for more than half of

41

the year. In early July he had not yet returned from his Spanish campaign, and his confidant Cornelius Balbus did not expect Caesar to arrive in Rome before the first of August (*Att.* 13.21a.3 of 30 June or 1 July). This means that if the *ludi Veneris Genetricis* were renamed in Caesar's honor and moved to 20-30 July in 45, as most scholars assume, the games were perversely held more than two months early, at a time when it was known that the honoree could not be present. Moreover, as the summer wore on, it became increasingly clear that Caesar expected to be back well before the end of September since he had assured Balbus that he would return before the *ludi Romani* of 4-18 Sept. (*Att.* 13.46.2 of 12 Aug.),[1] and he had indicated to one of the aediles charged with giving the games that he would attend the shows (*Att.* 13.45.1 of ca. 11 Aug.). A remark in one of Cicero's letters late in August implies that Caesar came back even a little ahead of this projected schedule ("our master is here sooner that we had expected": *magister adest citius quam putaramus*, *Fam.* 7.25.1 of ca. 23 or 24 Aug.), and the testimony of Suetonius (*Iul.* 83.1) about Caesar making his will on 13 September on his Labican estate puts Caesar back in the neighborhood of Rome well before the one-year anniversary of the games to Venus on 26 September. Finally, since the renaming of the month Quintilis in Caesar's honor was still in the future, and since Caesar had already provided the Roman populace with one major show in July 45 when he put on the *ludi Apollinares* through one of his city prefects (pp. 35-36), there was scarcely any incentive to move the *ludi Veneris Genetricis* from September to July in 45. Instead, there was every reason to celebrate those games once more in September, on the one-year anniversary of the dedication of Caesar's new temple to Venus Genetrix.

Added to these considerations is the fact that a celebration of the *ludi Veneris Genetricis* in September 45 would have been ideally suited to complement Caesar's triumphal return to Rome from his latest campaign (in Spain). It is to be remembered that at their inauguration in 46 B.C. those games had served to cap the celebration of Caesar's four triumphs. (It is, in fact, this double connection with triumphal returns of Caesar—in 46 *and* in 45—that most probably explains why the festival was later transformed into Victory games under the empire.[2]) In 45 Caesar celebrated his Spanish

---

[1] On the date of these games, see above, p. 31.21.

[2] Sulla had set the precedent by establishing *ludi Victoriae* (26 Oct. - 1 Nov.) to commemorate his victory in the Battle of the Colline Gate (1 Nov. 82). (Their name was later expanded to *ludi Victoriae Sullanae* so as to distinguish those games from the ones in Caesar's honor: see Wissowa 456.) One has to wonder, however, whether Caesar is likely to have followed this precedent, recalling as it did Sulla's proscriptions and the suppression of the Marians, with whom Caesar was connected. Caesar's aim more often

triumph very close to the one-year anniversary of the games to Venus.[3] The exact date of the triumph, unfortunately, is not attested because there is a gap in the *Fasti Triumphales* between 54 and the latter part of 45. A *terminus post quem non*, however, is furnished by the recorded date of the triumph that Caesar permitted his lieutenant Quintus Fabius Maximus to celebrate on 13 October,[4] which according to Quintilian (6.3.61) followed Caesar's own triumph by "a few days" (*post dies paucos*).[5] Caesar's triumph will have taken place, therefore, precisely in the period during which the games to Venus had been celebrated in 46 since those games presumably began on 26 September and ran at least through the end of the month, and probably into October.[6]

Finally, in 45 B.C. not only was the timing ideal for combining once more the games in honor of Venus Genetrix with a celebration of the achievements of her illustrious descendant, but we also happen to be informed that "lavish shows" (*regia munera*) were being planned during the summer of 45 (preparations attested by *Fam.* 6.19.2 at the end of July, if our revised dating is accepted, and by *Att.* 13.46.2 of 12 Aug.). These were the shows in which Cicero's friend Lepta hoped to play a role as "superintendent" (*curator*) and anxiously sought this commission through negotiations with Caesar's agents Balbus and Oppius (see p. 40). The involvement of Balbus and Oppius clearly demonstrates that Caesar had an interest in those entertainments, and it is logical to assume that the *munera* formed part of the preparations that were being made to celebrate Caesar's Spanish triumph in late September, early October 45.[7] We know that

---

than not appears to have been deliberately anti-Sullan, as for instance in his demolition of the Curia Hostilia, the old senate-house, and his plans to replace it with the new Curia Julia, since the Curia Hostilia was associated with the memory of Sulla: enlarged by Sulla in 80 and rebuilt by his son Faustus after the fire in 52 (Dio 44.5.2): see Weinstock (1957), 230-31. Better to assume that the games were renamed *ludi Victoriae Caesaris* by Caesar's successors, the triumvirs, who were less squeamish about adopting Sulla's tactics against their foes.

[3] Livy, *Per.* 116; Vell. 2.56.2-3; Suet. *Iul.* 37.1; Pliny, *NH* 14.97; Quint. 6.3.61; Plut. *Caes.* 56.3-4; Flor. 2.13.88-89; Dio 43.42.1-2.

[4] *Inscr. Ital.* XIII.1 87.

[5] Caesar's Spanish triumph is generally assigned to October (so *MRR* II 305), but the only evidence for this date is Quintilian's remark and a rather loose statement in Velleius (2.56.3) that Caesar lived only five months "after he had returned to Rome in October" (*cum mense Octobri in urbem revertisset*). Velleius' assertion, however may well leave out of account the few days at the end of September (26-30) during which Caesar could have attended the *ludi Veneris Genetricis* and brought them to a magnificent conclusion by his triumphant entry into Rome at the beginning of October.

[6] For the probable date of the *ludi Veneris Genetricis* in 46 and their relationship to Caesar's triumphs in that year, see Appendix III.

[7] So, for instance, Taylor 233 concludes. Lepta's bid to serve as a *curator* in 45

gladiatorial displays (*munera*) took place at the time of the *ludi Veneris Genetricis* in 46 and again at Octavian's games in 44. All these circumstances point inexorably to the conclusion that in 45 B.C. the *ludi Veneris Genetricis* were held in the same month as the year before, September, where an entirely suitable context can be found for them.

The Arguments for Moving the Games in 44 B.C.

In contrast with the previous year, 44 B.C. offered every incentive for holding games in Caesar's honor in July and for Octavian to combine with those games a proleptic celebration of the *ludi Veneris Genetricis*, apparently on the pretext that the board charged with sponsoring the festival was failing to act for one reason or another. July, of course, was the month in which Caesar had been born, and early in 44 an official act had changed the name of the month from Quintilis to Iulius in Caesar's honor. These facts alone provided powerful incentives for Octavian to select July as the month for holding the games he had vowed in May (Text 18), but an even greater incentive was provided by his desire to counter the bid for popular support that was being made by Marcus Brutus, one of the leading conspirators. It was the duty of Brutus, as urban praetor, to hold the *ludi Apollinares* on 6-13 July, and Brutus and his fellow conspirators pinned great hope on using those games to reverse the ill will that had spread among the urban populace and forced Caesar's assassins to withdraw from Rome in April for the sake of their personal safety (Becht 46-48). Therefore, the popular reception of Brutus' games in July was of paramount importance both to those who wished to suppress the memory of Caesar and to those who wished to keep it alive. Cicero, who remained distant from the city on his estates in Campania, clearly took an interest in the outcome of the bid for popular favor since he asked Atticus to send him detailed reports concerning Brutus' games and possibly also concerning those of Octavian that were to follow. [8]

Brutus spared no expense in producing his *Apollinares*, although he

---

reminds us of the role played by Matius and two other friends of Caesar in helping Octavian put on his shows in 44 (Texts 16, 18-19). It would not be surprising if the *collegium* that Caesar appointed in 46 to take charge of the *ludi Veneris Genetricis* enlisted in 45 the support of such free-lance agents as Lepta.

[8] Cicero asked to be told how Brutus' games were received "from their beginning" (*ab ipsa commissione, Att.* 15.26.1), and then he wanted daily accounts on "all the remaining games" (*omnia reliquorum ludorum*), which Shackleton Bailey (ad loc.) interprets as a reference to Octavian's *ludi Victoriae Caesaris*, although *reliqui ludi* could also be understood as referring solely to the rest of the days of the *Apollinares* after the opening day.

could not preside in person.[9] The hostility of the urban mob made it too dangerous for him to return to Rome from his self-imposed exile on his country estates. Cicero's friend Atticus seems to have taken a prominent role on Brutus' behalf, both in supplying money and in acting as his agent in Rome (*Att*. 15.18.2; cf. Nep. *Att*. 8.6). In July, Brutus had high hopes of achieving a sufficiently strong popular demonstration at the *Apollinares* to permit him to return to the capital (Appian 3.23.87). For this reason he delayed his departure from Italy until after the games to see if he could avoid having to take up the unwelcome Asian grain commission that had been voted to him in June.[10] Brutus urged Cicero to do him the honor of attending his games,[11] and during the first few days of July Brutus anxiously awaited news of the crowd's reaction, welcoming Atticus' detailed account received about 10 July (*Att*. 16.2.3).

Brutus' opponents appear to have been at work behind the scenes. Accius' play the *Tereus* was substituted for his *Brutus*, which the liberator had requested, recalling as it did his namesake who had expelled Rome's last king, Tarquinius Superbus.[12] The announcement of the date for the games incorporated the expression "Nones of July", a deliberate affront to the Liberators, who rejected the renaming of the Quintilis in Caesar's honor and regarded the day on which the games began (6 July) as *pr. Non. Quint*. (*Att*. 16.1.1).[13] Brutus coun-

---

[9] Plutarch (*Brut*. 21.3) attests Brutus' efforts to secure the best actors and the trouble he took to acquire a large number of wild beasts for his shows. In early June, Brutus had still hoped to be able to preside at the games in person, but he was talked out of this intention by his friends on the grounds that it was too dangerous for him to appear in Rome (*Att*. 15.11.2 of ca. 7 June, cf. ep. 12.1).

[10] Brutus was not expected to sail until after he had received reports at the conclusion of the games (*Att*. 16.4.4), and he had only recently set out with his fleet when Cic. crossed his path at Velia in southern Italy on 17 Aug. (*Att*. 16.7.5). The *curatio frumenti*, which charged Brutus with sailing to Asia, was viewed as a pretext cooked up by Antony to get Brutus out of the way, and it was a lowly task (*Att*. 15.10). See below, however, p. 102.24 for the grounds on which it may have been justified.

[11] *Att*. 15.26.1. The request put Cicero in an awkward position because he had not set foot in Rome for the past three months out of displeasure with Antony's regime, and Cicero could not bring himself to do so in July. Ordinarily a public figure of Cicero's stature would be expected to attend the games out of respect for the giver of the show. So, for instance, Cicero attended Pompey's games in 55, even though they were not to his taste (*Fam*. 7.1.1-2), the *Apollinares* in 54 out of regard for Fonteius (*Att*. 4.15.6), and Caesar's games in 46 (*Fam*. 12.18.2).

[12] *Att*. 16.5.1. The *Tereus*, however, apparently received a reasonably favorable reception (*Att*. 16.2.3, cf. *Phil*. 1.36, 2.31, 10.8). In contrast, the Greek shows seem to have been poorly attended (*Att*. 16.5.1).

[13] Presumably Brutus' surrogate, Mark Antony's brother, the praetor Gaius Antonius, employed the new name "Iulius" in the announcement (Appian 3.23.87; cf. Dio 47.20.2: see also Shackleton Bailey on *Att*. 16.2.3). The reference to the "Nones of July" (7 July) in *Att*. 16.1.1 and 4.1 doubtless lies behind Scullard's false assertion (160) that

tered the offensive "Nones of July" by insuring that the "hunt" (*venatio*), which he had arranged to hold on the 14th, was announced for the "14th Quintilis" (*Att.* 16.4.4). [14] The *venatio*, which is an attested feature of the *Apollinares* in the late republic, [15] apparently was regarded as something of an appendage and so had no fixed date. We know of at least one other occasion when it was held after the conclusion of the games on the 13th. [16] Brutus seems to have taken advantage of this flexibility by planning from the start to hold the hunt on the 14th, thereby currying popular favor by adding a second day of *ludi in circo*. [17] These strenuous efforts of Brutus to win popular support appear to have been offset to a certain extent by Octavian's distributions of money at the time of *Apollinares* (Appian 3.23.88), [18] and by counterdemonstrations which may have been organized to quell outbursts of sympathy for Brutus during the course of the games themselves (Appian 3.24.90-91). [19]

When Octavian and his advisors laid their plans in the spring of 44, [20] it should come as no surprise that they selected July as the month in which to hold games honoring Caesar. Octavian will have wanted to counter at the first opportunity any success that Brutus might achieve in turning the popular tide in his favor. Even the announcement of Octavian's games (in early May) and the preparations for them will have served to divert some of the attention away from Brutus' efforts to gain the limelight. [21] The fact that July was the

the *Apollinares* lasted only seven days in 44 (i.e., 7-13 July), although he, along with most scholars (e.g., Mommsen, *CIL* I² 321; Weinstock 156) accepts 8 days as their usual length (i.e, 6-13 July).

[14] Shackleton Bailey's emendation "*II Idus Quintilis*" (14 July) for the paradosis "*III Idus Quintilis*" (13 July) is plainly wanted since the same sentence explicitly states that the hunt was to take place on "the day following the *ludi Apollinares*" (*postridie ludos Apollinaris*), which ended on 13 July. Ville 106-108, without referring to SB, makes the case for this emendation, and it is adopted by Beaujeu in his Budé ed. of Cicero's letters, vol. 9 (1988).

[15] First attested at Sulla's *ludi Apollinares* in 93 (Pliny, *NH* 8.53; Sen. *De brev. vit.* 13.6); then in 54 (*Att.* 4.15.6) and in 41 (Dio 48.33.4).

[16] In 54: "the hunt has been postponed until later" (*venatio in aliud tempus dilata, Att.* 4.15.6).

[17] The *Apollinares* were also extended to include one extra day of *ludi in circo* by Agrippa in 40 when the *lusus Troiae* was performed (Dio 48.20.2). Cf. the addition of one day to the *Plebeii* [?] in 56 (*Att.* 4.8a.1).

[18] Taylor, *Divinity* 90 speculated that this may have been on 13 July, Caesar's birthday.

[19] However, Cicero's failure to mention these counterdemonstrations should cause us to be wary of accepting Appian's account at face value: see Ehrenwirth 50 n. 6.

[20] Chief among these advisors were doubtless Balbus and Oppius, to whose importance Alföldi (1976) 31-54 has directed attention.

[21] Preparations for shows or banquets could be used to impress upon the public the splendor of the coming event. So, for instance, Julius Caesar as aedile in 65 displayed in temporary colonnades the equipment that was to be used in his shows (Suet. *Iul.* 10.1),

month of Caesar's birth and the name of the month had earlier in the year been changed to "Iulius" were clearly factors that Octavian must have taken into account. In fact, all of the arguments advanced by Weinstock (156) for selecting July as the month in which to hold games honoring Caesar have greater weight if the year in which the games were first held was 44, rather than 45 as Weinstock believed.

Once Octavian had made the decision to hold his games for Caesar in July, it remained to decide upon a date. The range of dates available to Octavian in that month was quite narrow but nonetheless highly satisfactory, as chance would have it. The *ludi Apollinares* occupied 6-13 July, and the next six days (14-19 July) were given over to a "market" (*mercatus*),[22] one day of which Brutus had already appropriated for his *venatio* on the 14th. Therefore, the 20th was, by default, the closest date to Caesar's birthday (13 July) on which Octavian could possibly hold his games and still leave some space between his shows and those of Brutus. The market days probably assured that the crowd drawn by the *Apollinares* would not drift away after the conclusion of the first set of games, although it is difficult to gauge how functional (as opposed to being a fossilized institution) the *mercatus* may still have been in the late republic. Finally, 20 July (*a. d. XIII Kal. Sext.*) will have been exactly four months from the date of Caesar's funeral, which scholars generally agree was held on 20 March (*a. d. XIII Kal. Apr.*),[23] a day made memorable by the burning of Caesar's body on a makeshift pyre in the Forum and riots in the streets of Rome.[24]

---

and in 46 he had Roman knights and senators take into their homes and train the gladiators that were to be displayed in his *ludi Veneris Genetricis* (Suet. *Iul.* 26.2). The competition for resources in 44 (actors and wild animals) doubtless explains Brutus' instructions to his agents to use up all of the beasts in the *venatio* and not to put any on sale (Plut. *Brut.* 21.3). Otherwise they might fall into the hands of Octavian.

[22] Degrassi 482-83; this feature is attested for only two other *ludi publici*, the *Romani* and *Plebeii*, followed respectively by a *mercatus* on 20-23 Sept. and 18-20 Nov.: see Wissowa 454.

[23] Becht 84. Furthermore, if the games in 44 occupied 20-28 July, as we shall argue below on other grounds (see pp. 54-55), the festival will have lasted for exactly nine days, the traditional length of the solemnities in a Roman funeral ("hence the games celebrated in honor of the dead are called *novendiales* [i.e., lasting nine days]": *inde etiam ludi qui in honorem mortuorum celebrabantur novendiales dicuntur*, Porphyrion on Hor. *Epod.* 17.48).

[24] Arguably the date of most Roman funerals would soon be forgotten, but Caesar's was exceptional and the date must have been seared in the public's memory. The armed assaults on the houses of the conspirators, the burning of the house of the senator Bellienus (*Phil.* 2.91), and the tragic death of pro-Caesarian tribune Helvius Cinna, who was mistaken for the praetor Cornelius Cinna and torn to pieces by an angry mob (Plut. *Caes.* 68.2-3, *Brut.* 20.5-6; Suet. *Iul.* 85; Dio 44.50.4; Appian 2.147.613; Val. Max. 9.9.1), insured that the day would be long remembered. So too, the impromptu burning of Publius Clodius' body in the Curia in 52 B.C. caused the date of that event, 19 Jan., to find its way into the historical record: Ascon. pp. 32-33 (Clark).

Nature of the Celebration in 44 B.C.

This last point is a potentially significant one to keep in mind as we turn next to consider the nomenclature and nature of the games given by Octavian in July 44. Although Mommsen (181) tersely dismissed as unreliable the notices in Servius that describe Octavian's games as "funeral games" (*ludi funebres*, Texts 5-8), there is no good reason for doubting those statements.[25] It would be quite natural for Caesar's adopted son to hold funeral games as an act of filial piety, *pietas* being a concept that became a watchword with Octavian. This way of honoring the dead was an accepted Roman practice, and it would have been entirely appropriate for Octavian to complement his act of formally accepting his inheritance from Caesar by vowing to hold such games.[26] As we have just observed, one reason why Octavian selected 20 July for holding his games may have been that the date fell exactly four months after the day of Caesar's funeral on 20 March. Furthermore, if scholars are correct in viewing the games of 44 and the appearance of the comet as in part inspiring Virgil's description of the funeral games for Anchises and the portentous fiery arrow of Acestes in the *Aeneid* (5.519-44),[27] then we have

---

[25] Williams 142 also rejected Servius' characterization of the games as *ludi funebres*, but in one passage (Text 5) Servius makes it clear that the funeral games were merely combined with a celebration of the *ludi publici* in 44, and this reconstruction of Octavian's games is adopted by, among others, Taylor, *Divinity* 89, Weinstock 89, 368, and Shackleton Bailey on *Fam.* 11.27.7. Since the *ludi funebres* will inevitably have included *munera*, the gladiatorial displays in 44 should be reckoned among the eight *munera gladiatoria* that Augustus claimed to have given (*RG* 22.1). Scholars have tended to overlook the games in 44, stating that we can identify only seven of the eight occasions when such shows were given (so, e.g., among editors of the *RG*: Gagé 118-19; Volkmann 39; Brunt-Moore 64). Weber 230 correctly takes into account the games in 44, while Ville 122-23 offers a different, tentative identification of the eight occasions, without including the games in 44.

[26] There was no constraint of time, however, and so the choice of July was purely arbitrary and motivated no doubt by the considerations discussed above. Years could elapse between a death and funeral games in honor of the deceased: e.g., funeral games for Julius Caesar's father (d. ca. 85 B.C.) in 65 (Suet. *Iul.* 10.2; Dio 37.8.1), for Caesar's daughter Julia (d. 54 B.C.) in 46 (see p. 52.39), for Metellus Pius (d. ca. 63 B.C.) in mid 57 (Cic. *Sest.* 124), and for the dictator Sulla (d. 78 B.C.) in 60 (*Vat.* 32; Dio 37.51.4; cf. *Sull.* 54-55).

[27] A view originating, it seems, with Wagner 244 and taken up by Drew 43-47, Wagenvoort 26-27, and Bömer 31. (Williams 142, following Heinze 165-69, rejects this interpretation, but see now West 9-13 for a cogent refutation of Heinze.) Henry 132 draws attention to the similarity between the account of Acestes' arrow catching fire and Ovid's account of the metamorphosis of Caesar's soul into a fiery comet (Text 30). Another possible point of resemblance between Octavian's funeral games for Caesar and Aeneas' funeral games for Anchises may have been a performance of the horse riding pageant known as the *lusus Troiae*, which Virgil credits with first being performed at the

additional indirect evidence that Octavian's games were at least partly *ludi funebres*.

The evidence furnished by Matius' letter (*Fam.* 11.28.6), which was written shortly after Octavian's games in 44,[28] also tends to support the view that the celebration included *ludi funebres* for Caesar, just as Servius asserts. Weinstock (369) has rightly pointed out that the way in which Matius justified his conduct in serving as one of Octavian's "agents" (*procuratores*) who took responsibility for producing the games suggests that Matius viewed his role chiefly in the light of making possible the funeral games. Matius wrote of the "duty" (*munus*) that he was obliged to shoulder "for the memory and honor of a dear, departed friend" (*hominis amicissimi memoriae . . . etiam mortui*, Text 16), and the word *munus*, which may denote a service performed or a gift given out of kindness or duty,[29] can refer more particularly to the tribute or offering that is owed to the dead (*OLD* s.v. 3). Matius' words are echoed in Cicero's observation that Matius acted "from a sense of obligation and human decency" (*pie et humane*) in his "superintendency of the games" (*curatio ludorum*, Text 19).[30] As we argued earlier in chapter one (p. 5), the emphasis that Matius' letter appears to have placed on the *ludi funebres*—in contrast with our other sources which tend to single out the games in honor of Venus Genetrix—may explain why Matius referred to the games as those "which young Caesar [Octavian] celebrated in honor of [Julius] Caesar's victory" (*quos Caesaris victoriae Caesar adulescens fecit*, Text 16). The words *Caesaris victoriae* certainly recall the name by which the festival was known in imperial times (*ludi Victoriae Caesaris*), but given the slightly different word order (in part influenced, no doubt, by the wish to avoid the collocation

---

games for Anchises (*Aen.* 5.548-603). Although not directly attested for Octavian's games in 44, the *lusus Troiae* is likely to have been performed, to judge from the part it played in the *ludi Veneris Genetricis* of 46 (Dio 43.23.6; Suet. *Iul.* 39.2), from its great popularity under Augustus' regime (Suet. *Aug.* 43.2), and finally from the fact that Baebius Macer may have touched upon the *lusus Troiae* in his account of Octavian's games (see p. 84.61).

[28] For the date of this letter, see p. 4.11. The relevant passage is Text 16.

[29] "*Munus* has the meaning *officium* [a service done out of kindness or duty] when a person is said to perform a *munus*. Likewise it can refer to a gift that is given because of an *officium*" (*munus significat <officium>, cum dicitur quis munere fungi. Item donum quod officii causa datur.* Paulus *ex* Festus p. 125 Lindsay).

[30] There is no difficulty in interpreting the term *ludi* as a reference to *ludi funebres* since we know a distinction in terminology was customarily maintained between the "shows" (*ludi*) and the display of gladiators (*munera*), which had become a traditional way of honoring the dead: e.g., Livy 28.21.10, "funeral games were added to this gladiatorial display" (*huic gladiatorum spectaculo ludi funebres additi*), and 31.50.4, "funeral games were held, and a show of gladiators was given" (*ludi funebres . . . facti et munus gladiatorium datum*); cf. 23.30.15 and 39.46.2.

Caesaris Caesar), the expression in Matius' letter may best be interpreted as a descriptive phrase, referring principally to the *ludi funebres* rather than to the public festival to which the funeral games were attached.

Clearly, however, there was another important element of Octavian's games that Matius may have refrained from mentioning directly because it was not his strongest debating point in refuting the criticism that he was showing excessive devotion to the memory of Julius Caesar. As we have seen, *all* of the sources that name the public festival during which the comet appeared (Texts 1-5)—as well as two of the sources that mention only the *ludi*, but not the comet, (Texts 14, 15)—identify those games with the festival that Caesar had founded (in 46) in honor of Venus Genetrix. There is also a tradition that in 44 Octavian assumed the responsibility for giving the games to Venus because the "board" (*collegium*) that was responsible for holding the *ludi* either neglected their duty (Text 3) or were afraid to carry it out (Text 17). Both explanations sound like pretexts for Octavian to take matters into his own hands, the latter justification putting the conduct of the board in a more favorable light.[31] Octavian, therefore, acted "on behalf of the board" (*pro collegio*, Text 4), and Pliny (Text 1A) adds the detail that Octavian was himself "a member of the board that had been established by Julius Caesar" (*in collegio ab eo* [sc. *Iulio Caesare*] *instituto*). Pliny's statement should not be dismissed lightly.[32] Octavian's prominent role in presiding over the Greek theater at the first celebration of the *ludi Veneris Genetricis* in 46 (Nic. Dam. 9.19) points to the conclusion that Octavian was in all probability appointed to the board by his great-uncle at that time.[33] That *collegi-*

---

[31] The extent to which the complaint against the board was most likely special pleading (doubtless a theme in Octavian's propaganda against Mark Antony: see Levi I 108) can be judged from Dio's remark (Text 3) that those who had promised to hold the games in Caesar's honor at the *Parilia* (on 21 April) also took their responsibilities lightly. We happen to know, however, that circensian games honoring Caesar were presented at the *Parilia* in 44, on what seems to be quite a respectable scale (*Att.* 14.14.1, ep. 19.3). See Alföldi (1953) 47-48 for the evidence of the coinage: Antony with beard of mourning (ob.), "circus horseman" (*desultor*) (rev.); cf. Crawford I 495.

[32] Treated by Mommsen (*CIL* I² 322) as an error ("*minus accurate*"), but accepted by Drumann-Groebe I 91 and Taylor, *Divinity* 63. Since Pliny's remark introduces an extended quotation from Augustus' *Memoirs* concerning the comet and games (Text 1), Pliny in all likelihood drew upon that source for his description of the role played by Augustus.

[33] Octavian could also have played a role at the celebration of 45, if it took place, as we have argued, in Sept. (more precisely 26 Sept.–early Oct.: see Appendix III), since Octavian had returned to Rome by then and did not leave for Apollonia until Oct. at the earliest (Appian 3.9.32) or more probably Dec. (Nic. Dam. 16.37-38). See Drumann-Groebe 425 for a defense of the chronology in Nicolaus of Damascus; cf. Fitzler-Seeck, *RE* 10.1 (1917) 279.

um presumably had close ties to the *gens Julia* since it oversaw the festival to "Venus the Ancestress [of the Julian *gens*]",[34] and it makes sense for Caesar to have included Octavian on the board in 46 B.C., particularly if Caesar was perhaps already thinking of Octavian as a potential heir (in advance of his formal adoption by will one year later). To act *pro collegio* (Text 4) does not exclude the possibility that one is a member of the *collegium* (Text 1A) as we can tell from what appears to be a comparable use of the expression *pro collegio* in Augustus' *Res Gestae* (22.2). In his account of the *ludi Saeculares* of 17 B.C., Augustus states that he gave the games "on behalf of the board of Fifteen, as its chief officer" (*pro collegio XV virorum magister collegi*, cf. *RG* 7.3).[35]

Octavian's reasons for assuming the right and responsibility for giving the games to Venus are not hard to imagine. Although measures had been passed in 45 and 44 that tended to elevate Julius Caesar to the status of a god in his own lifetime,[36] and although an attempt had been made in April to establish an altar in the Forum on the spot where Caesar's body had been cremated, the consuls strenuously resisted the movement to recognize Caesar's divinity in the months that followed his death.[37] Even before the games in July, Octavian had tried and been rebuffed in his attempt to display two symbols of Caesar's elevated status, his golden chair and bejeweled crown. A decree passed in early 44 had provided for this crown and chair to be set up in the theater at all the games,[38] but Antony forbade Octavian to display those tokens of Caesar's superhuman status even at the games that

---

[34] Symmachus (*Laud. in Valent. sen.* 2.32) alludes to this close connection: "the Julian house looks after the sacred rites of Venus" (*Veneriis sacris famulata est domus Iulia*). This must be the basis of Kornemann's assertion (*RE* 4 [1901] s.v. "collegium", 384) that the *collegium* appointed by Caesar in 46 comprised members of the Julian *gens*.

[35] In *RG* 22.2, *pro* ("on behalf of") clearly means "as their representative" (*OLD* s.v. 3). Therefore, this may also be the meaning of *pro collegio* in Obsequens' compressed account of Octavian's action in 44 (Text 4), despite the fact that Dio's account (Text 3) and Suetonius' (Text 17), taken on their own, might lead us to surmise that Octavian was not himself a member of the board.

[36] Scholars generally assume that formal divinization did not take place until 42 B.C. (see p. 53.43), but in the *Second Philippic* (2.110) of late 44 B.C., Cicero bitterly refers to the vote taken earlier in the year that granted the "deified Julius" (*divus Iulius*) a "priest" (*flamen*) and a whole host of other divine honors.

[37] The altar was erected in late March by a faction organized by Amatius, who had taken the name Gaius Marius and claimed to be the grandson of the general Marius. Amatius was put to death by Antony in early April (*Phil.* 1.5; *Att.* 14.8.1), and the altar was pulled down by Antony's colleague Dolabella later in the month (*Phil.* 1.5, 2.107; *Att.* 14.15.1).

[38] Dio 44.6.3; cf. *Att.* 15.3.2; Plut *Ant.* 16. For a discussion of the date of the decree and the significance of the symbols, see Weinstock 281-83.

Octavian was intending to give at his own expense (Texts 14, 15; cf. Text 3). Under these circumstances, Octavian appears to have adopted the next best course. If he could not openly advertise Caesar's divinity, he could at least indirectly convey his message by using the *ludi Veneris Genetricis* to lend an aura of divine majesty to Caesar's funeral games. The close connection between Venus and her most famous descendant had, in fact, recently been emphasized at Caesar's funeral when his body was laid out in a gilded replica of the temple of Venus Genetrix that was placed on the rostra (Suet. *Iul*. 84.1). There was also a precedent for attaching funeral games for a member of the Julian *gens* to the *ludi Veneris Genetricis*. At their foundation in 46, Caesar had combined with those games *ludi funebres* for his daughter Julia.[39] Octavian's act of linking the funeral games for Julius Caesar with a proleptic celebration of the *ludi Veneris Genetricis* (on the pretext that the *collegium* could not, or would not hold the festival later in the year, in Sept., Texts 3, 17) caused the games in 44 to have a two-fold nature. This circumstance will most easily explain why Octavian's games are variously described in our sources as either the *ludi Veneris Genetricis* (Texts 1-4, 14-15), or *ludi funebres* (Texts 6-8, supported indirectly by the testimony of Matius [Text 16] and *Aeneid* 5.519-44), or once (Text 5) more accurately, it seems, as a joint celebration in honor of Venus and Caesar.[40]

There is one additional piece of evidence, a passage in Suetonius (Text 9), that tends to point to the conclusion that when he celebrated his games in 44 Octavian broke new ground and laid the foundation for the imperial *ludi Victoriae Caesaris*. The relevance of this passage, however, has in the past been overlooked, doubtless because the transmission of the text has been called into question. As it is transmitted by the MSS, the text states that a comet appeared "at the games which Caesar's heir Augustus established in his honor and *celebrated for the first time* [emphasis added]" (*ludis, quos primo consecratos* [emphasis added] *ei heres Augustus edebat*).[41] In order to bring this statement into line with the prevailing modern view that the *ludi Victoriae Caesaris* had already in 45 B.C. been established in July, all modern

---

[39] Plut. *Caes.* 55.2; Dio 43.22.3; cf. Suet. *Iul.* 26.2; see Weinstock 89 for discussion.

[40] It is interesting to note that Gibbon IV 290, gave precisely this account of Octavian's games: "After the death of Caesar, a long-haired star was conspicuous to Rome and to the nations, during the games which were exhibited by the young Octavian *in honor of Venus and his uncle* [emphasis added]."

[41] That is, the participle *consecratos* ("established (as a holiday)", *TLL* IV.383.37ff) stands for a second verbal idea, whose subject is Octavian: = *quos ei heres Augustus tum primum consecravit edebatque*. We thank Robert Kaster for discussing this passage with us and helping us to clarify some of our thinking, without of course assigning to him any responsibility for the views expressed here.

editors adopt an emendation first printed in the Basel edition of 1546: *quos primo<s> consecrato[s] ei heres* . . . ("which Augustus first celebrated for Caesar after his consecration").[42] More than twenty years ago G. V. Sumner (291-92) defended the paradosis *consecratos* on the grounds that Caesar was not *consecratus* ("deified", "regarded as divine", *TLL* IV.383.45ff) until a year and a half later.[43] (It might also be pointed out that if we adopt *consecrato ei*, the pronoun is otiose.)

We now see, however, that there is potentially a far more cogent reason for retaining the paradosis. According to our reconstruction, when Octavian held his games in July 44, he could be viewed as laying the foundation for the festival that was later to be known as the *ludi Victoriae Caesaris*.[44] The games in 44, we believe, were indeed the first in Caesar's honor, just as Suetonius states they were in Text 9. In the one other passage where Suetonius refers to Octavian's games (without, however, mentioning the comet) he calls them the *ludi Victoriae Caesaris* and yet describes them in a way reminiscent of Dio's description of the *ludi Veneris Genetricis* in 44 (Text 3). This contradiction between the two passages in Suetonius is not unlike discrepancies that turn up elsewhere in the *De vita Caesarum*. Suetonius is

---

[42] So Ihm (ed. mai. 1907 and min. 1908), Rolfe (Loeb 1913), and Ailloud (Budé 1931). Butler-Cary ad loc., comment: "A necessary correction. . . . The games in question had been in existence since 46 B.C."

[43] The *communis opinio* is that Caesar was consecrated in 42 (modern scholarship summarized by Alföldi [1973] 229-30 & n. 8), despite Alföldi's attempt to prove that Caesar had already been made a god in his lifetime. However, as Robert Kaster has pointed out to us by letter (13 Nov. 1993), Suetonius may have jumped to the false conclusion that Caesar was already *consecratus* in 44 since the words introducing Suetonius' description of the comet (Text 9) state that Caesar "was assigned to the company of the gods" (*in deorum numerum relatus est*, cf. *ILS* 72, quoted below, p. 56.51) "not only by a decree" (*non ore modo decernentium*: i.e., the one attested by *ILS* 72) "but also by the conviction of the common people" (*sed persuasione vulgi*: explained as being fueled by the appearance of the comet). In that case, Sumner's argument loses its force. Even so, we can now approach the question from another direction. *Consecratos* may be defended on the grounds that the text transmitted by the MSS reflects historical fact: the games given by Octavian in 44 *were* the first to be "established" in Caesar's honor, as opposed to those honoring Venus Genetrix.

[44] We need not be surprised to discover that Augustus takes no credit in his *Res Gestae* for establishing this festival. That topic had clearly been treated already in Augustus' *Memoirs, De vita sua* (composed ca. 24 B.C.), which we know gave an account of the comet and other events surrounding the games (as attested by Texts 1 and 6). This material was not repeated in the *RG*, a later work (the extant version being a draft of ca. 2 B.C., with minor later revisions: Brunt-Moore 6). The *RG* also fails to mention, for instance, Octavian's inheritance from Caesar and his struggle to gain recognition of his testamentary adoption, subjects that were certainly treated in the *Memoirs* (Schmitthenner 12 and n. 1). For the different aims of the *Memoirs* and *RG*, see Yavetz (1984) 1-36.

quite capable of telling two different versions of the same anecdote to suit the needs of a given context, sometimes even within the same biography.[45] Furthermore, it is quite easy to understand how the discrepancy between the two passages concerning Octavian's games could have arisen if the celebration in 44 had the two-fold character that we have discussed, and if Suetonius, unlike the bulk of our sources, chose in his biography of Caesar (Text 9) to emphasize that the games in 44 included for the first time honors for the late Caesar, soon to be deified.

Scale of the Celebration in 44 and Later History

Finally a word is in order about the length of the celebration in 44 and the later history of the festival. As we have stated, the appearance of the comet and its interpretation as a sign of Caesar's apotheosis must have insured that the future celebrations of the games were held on the same dates as in 44. This is not to say, however, that Octavian's games in 44 necessarily equalled the length of the *ludi Victoriae Caesaris* in imperial times, eleven days (20-30 July), on four of which circensian games were held.[46] Two or three considerations point to the conclusion that the celebration lasted only eight or nine days, at the most, in 44. (1) Four days of *ludi in circo* must be regarded as extremely lavish and probably beyond Octavian's means in 44. In the late republic, the only other *ludi publici* to have more than one day of circensian games were the *Romani* and *Plebeii*, having four and two days respectively. (2) We can be fairly certain that Octavian's games lasted longer than seven days because the comet is said to have been visible for seven days during the course of the games (Texts 1, 9, 10, 12), but only one source (Text 3) mistakenly [?] equates the period during which the comet was seen with the

---

[45] Compare, for instance, the two versions of how Claudius died (by poisoned mushrooms, *Nero* 33.1; by a poisoned drink, *Nero* 39.3—possibly each drawing upon the two conflicting accounts reported in *Claud.* 44.2); the role of Tiberius in bringing about the death of Germanicus (Tiberius used Piso as his agent, *Tib.* 52.3; no mention of Tiberius' responsibility, *Calig.* 6.1-2); the burning of indictments (actually burned, *Calig.* 15.4; had not in fact been burned, *Calig.* 30.2); letters of Augustus on the character of Tiberius (positive, *Tib.* 21.3; negative, *Tib.* 51.1); Tiberius' refusal of title *pater patriae* (out of moderation and humility, *Tib.* 26.1-2; for less noble reasons, *Tib.* 67.2-3). We thank Keith Bradley for discussing with us Suetonius' capacity for admitting into different parts of his work a contradiction of the type that we detect between *Iul.* 88 and *Aug.* 10.1. We also thank John Burke for helping us to collect the above list of parallels in support of our thesis.

[46] Only a few scholars seem to have allowed for the possibility that the games in 44 did not occupy all eleven days: Schmidt (1883) 864, because he wanted to make room for Antony to pass a *lex de permutatione provinciarum* at the end of July, and Tyrrell-Purser on *Fam.* 11.28.6.

length of the festival as a whole. Furthermore, since the *ludi Apollinares* ran for eight days (6-13 July) and Octavian must have wanted to match their splendor, we may surmise that Octavian set aside at least an equal number of days for his games. (3) Finally, since Brutus increased the length of his *Apollinares* in 44 by arranging to hold the hunt on the "14th Quintilis" (*Att.* 16.4.1), it may not be too fanciful to interpret this decision as an effort on Brutus' part to compete with a prior announcement of Octavian that his games would include two days of circensian games.[47] A festival running from 20 to 28 July (9 days, cf. n. 47.23) would have been quite respectable in 44 and may be assumed to be in line with Octavian's aims and resources at that time.

As for the later history of this festival, whatever length and form it had in 44, it is extremely unlikely that it was repeated in July of the following year. Dio (36.31.4) tells us that in the spring of 43 the Roman treasury was so depleted that even the time-honored, annual festivals had to be curtailed. Cicero's letters in May and June bear out the notion that the public coffers were drained of cash, unable to meet the pressing need to pay the troops and make good the rewards promised to the victorious soldiers.[48] Then too, from their inception in 46, the games in honor of Venus Genetrix had been the responsibility of a board established by Caesar, rather than one of the elected magistrates of the state, and as we have seen, already in the previous year the board was charged with failing to shoulder its responsibilities either out of fear (Text 17) or negligence (Text 3). Presumably that board was no more energetic in 43 when, on top of all else, it would have been difficult to raise cash and assemble the resources needed to stage an elaborate festival. By July, when the one-year anniversary of Octavian's games rolled round, the troubled conditions that are attested for the spring could only have grown worse. Rome was in chaos, and one has to wonder whether the state could have met the obligation of celebrating even the venerable *ludi Apollinares*.[49] Before the month was out, Octavian sent to the Senate a delegation of 400 centurions and soldiers to demand the consulship, which had been left vacant by the deaths of Hirtius and Pansa in the struggle against Mark Antony in northern Italy.[50] When the Senate failed to meet the demands of the

---

[47] This is where we could find room for a performance of the *lusus Troiae*: see above, p. 49.27.

[48] Attested in the exchange of letters with Decimus Brutus in the north of Italy and Quintus Cornificius, governor of Africa: *Fam.* 11.10.5, ep. 24.2, ep. 26; 12.30.4.

[49] Marcus Caecilius Cornutus, the praetor urbanus whose duty it was to put on the *Apollinares*, had his hands full in 43 since the day-to-day management of affairs fell to him after both consuls perished in the war at Mutina (*MRR* 2.338).

[50] *RE* 7A (1939) 1083. Syme 185-86 nicely captures the mood of those troubled weeks.

nineteen-year-old Octavian, he marched with his troops against the capital and caused himself to be elected consul on August 19 (*MRR* 2.336). Under such conditions, we can hardly imagine that the new festival—whether it be called the *ludi Veneris Genetricis* or *Victoriae Caesaris*—was celebrated in July 43. This point is worth bearing in mind if, contrary to what most scholars believe, the festival had not already assumed its final form in 45 but underwent in 44 significant changes in date and content, as we have argued. The hiatus in 43 must have allowed further flexibility to whoever revived the festival and made it a regularly recurring celebration in July.

Presumably the next occasion on which the games were held was in 42, under the regime of the triumvirs. The triumvirs had in the meantime caused Caesar's divinity to be recognized by an act of the Senate and Roman people[51] and had laid the foundation for a temple in Caesar's honor on the spot in the Forum where his body had been cremated (Dio 47.18.3; cf. Appian 2.148.616-617). Since Antony and Octavian were engaged overseas in the military campaign against Brutus and Cassius, which was to culminate in the final battle at Philippi on 23 October, the responsibility for seeing that the games were held in July must have fallen to the remaining triumvir Lepidus, perhaps with the cooperation of his colleague in the consulship Lucius Munatius Plancus.[52] If we did not happen to have evidence to the contrary, it would be tempting to suppose that the *ludi Veneris Genetricis* were renamed in Caesar's honor at the same time that the other measures recognizing Caesar's divinity and providing for his worship were adopted in 42. We have, however, an attested celebration of the *ludi Veneris Genetricis* in 34 B.C. by the consuls (Dio 49.42.1). Modern scholars have tended to brush aside the reference to Venus Genetrix since it scarcely fits their view that the name of the games had already been changed to *ludi Victoriae Caesaris* more than ten years earlier in 45.[53] The apparent anomaly is more easily explained under our reconstruction of the festival's history which views the games in 44 as the first occasion on which the games dedicated to Venus Genetrix were expanded to include honors for Caesar. Also, if the *ludi Veneris Genetricis* had been

---

[51] Attested by *ILS* 72: "to the spirit of the deified Julius, father of this country, whom the Senate and Roman people assigned to the company of the gods" (*Genio deivi Iuli parentis patriae, quem senatus populusque Romanus in deorum numerum rettulit*).

[52] If so, that occasion might provide the precedent for assigning the games to the care of the consuls, as we know they were in 34 B.C. (discussed immediately below).

[53] Wissowa 457, for instance, does not quote the relevant passage in Dio but simply cites it as evidence that the *ludi Victoriae Caesaris* had been placed in the hands of the consuls. Dio, however, writes that "the consuls [of 34 B.C.] celebrated the festival that was consecrated to Venus Genetrix" (τὴν πανήγυριν τὴν τῇ Ἀφροδίτῃ τῇ γενεθλίῳ τελουμένην οἱ ὕπατοι ἐποίησαν).

held not just the one time in 46 B.C. (as scholars generally believe) but in two successive years (46 and 45) as a companion piece to Caesar's triumphs, it is easier to understand how the games eventually were converted to Victory games when Divus Iulius assumed the dominant role.

Two final pieces of evidence are worth mentioning because they point to the same conclusion that Venus was not replaced by "Victoria Caesaris" as early as scholars have wanted to believe. (1) Among the acts of the triumvirs that were passed in 42 B.C. to promote the worship of Caesar, we are told that they arranged for his statue to be carried with the statue of Venus in the *pompae* (Dio 47.18.4). This is surely a significant detail in view of the likely dual nature of the games in 44. It suggests that Venus and Caesar continued to share the festival when it was revived in 42. (2) In the charter of the Colonia Genetiva Iulia Ursonensis in Spain, we find a provision for the celebration of annual games in honor of Venus, undoubtedly on the model of what was being done in Rome.[54] Since this colony was founded in the period that concerns us—its charter was issued by Mark Antony in 44, "at the behest of the dictator Gaius Caesar" (*iussu C. Caesaris dictatoris,* § 106)—we should not be surprised to learn, as we have, that the *ludi Veneris Genetricis* were undoubtedly still being celebrated in Rome, long after modern scholarly opinion would have those games thoroughly transformed into the *ludi Victoriae Caesaris*. The process of modification turns out to have been much more gradual than has hitherto been suspected. Also, we should now grant to Octavian the credit that is his due, since all of the evidence points to the conclusion that the *ludi Veneris Genetricis* had not already been moved to July and renamed in 45 B.C. Octavian turns out to be the one who first held the games in July and made room for Caesar to join the goddess Venus. This first public act on the part of the future Roman emperor foreshadows his clever use of symbolism and ceremony on many later occasions when he sought to promote his personal interests and those of his family.[55]

---

[54] *ILS* 6087.71: "those who become aediles, during the course of their magistracy, are to perform the duty of celebrating stage shows lasting one day in honor of Venus either in the circus or in the forum" (*Aediles quicumque erunt in suo magistratu munus ludos scaenicos . . . unum diem in circo aut in foro Veneri faciunto*).

[55] Compare, for instance, the games he celebrated in 29 B.C., following his triple triumph and dedication of his temple to Divus Iulius, (Dio 51.21-22) and the *ludi Martiales* instituted in 2 B.C. at the dedication of his temple to Mars Ultor (vowed before the Battle of Philippi in 42) and made an annual festival (*RG* 22.2). See also Zanker 168-70 on Augustus' staging of the *ludi Saeculares*.

PART II

CAESAR'S COMET

IV

THE ANCIENT ACCOUNTS OF THE COMET

OF 44 B.C.

THE HISTORICAL REALITY OF CAESAR'S COMET

Many sources (Texts 1-13) report that a distinctive, daylight comet (or new "star") appeared in the sky during seven of the days on which Octavian celebrated his games in 44 B.C. According to the bulk of those sources, the omen was interpreted as a sign of Caesar's apotheosis, quite an extraordinary interpretation to be given a comet. Ordinarily comets were regarded as baleful omens and aroused feelings of apprehension and dread.[1] In keeping with this tradition, Shakespeare has Caesar's wife Calpurnia exclaim on the morning of the Ides of March in 44, when she expresses her forebodings:

> When beggars die, there are no comets seen;
> The heavens themselves blaze forth the death of princes.
> *(Julius Caesar*, act II, scene 2.30-1)

The coincidence of a bright comet with such a momentous event as Caesar's funeral games, and the pivotal role that the comet is said to have played in causing the Roman people to recognize Caesar's divinity, cannot help but arouse suspicion. Scholars have noted that there is a tendency, particularly on the part of our western sources, to alter the dates of comets so as to assimilate them to great historical events.[2] Could this explanation

---

[1] See p. 135.2-4. In chapter seven we investigate the evolution of the positive interpretation of Caesar's comet and try to explain why it came to be regarded as such a favorable omen.

[2] Barrett 82-3; Schove iv-v, xxxii-xxxiv.

account for the comet of 44? Was the so-called Julian star (*sidus Iulium*) a product of invention, or was it perhaps the misrepresentation of an earlier, or later, historical comet that became attracted to the year 44? Certainly the mood of the people in Rome after the Ides of March was ideally suited for imaginations to run riot and conjure up all sorts of strange and unusual manifestations of the gods' displeasure. As Machiavelli remarked in his *Discorsi* (214), "Both modern and ancient examples go to show that great events never happen in any town or country without their having been announced by portents, revelations, prodigious events or other celestial signs." Indeed, the comet of 44 is sometimes treated as an omen with sinister implications, more closely allied to Caesar's demise than to his apotheosis (Texts 3, 6, 10-12). Concerning the period following Caesar' murder, Virgil goes so far as to assert that "at no other time did fearsome comets so often blaze" (*non alias . . .diri totiens arsere cometae*, Text 31).

Can it be that such a belief in the inevitability of omens ultimately lies behind the report in our sources that a comet made its appearance during the games held by Octavian in July? What a remarkable coincidence it must have seemed to contemporary observers, if on the occasion of Caesar's funeral games a bright light suddenly appeared in the sky, so bright that it could be seen well over an hour before sunset (Texts 1, 2, 4, 5, and 9)! If, as most scholars believe, the festival to Venus Genetrix had already been moved to July and rededicated to Victoria Caesaris in the previous year (45 B.C.), then the concurrence of the games and comet in July 44 would have been striking enough. But if, as we have argued, Octavian deserves the credit for moving the festival to July in 44 and for giving a share of the honors to Caesar, then the sudden appearance of the comet during those seven days must have seemed to those who witnessed it nothing short of a miracle from heaven.[3]

---

[3] This is an appropriate point to remind ourselves that truly amazing coincidences do sometimes happen. For instance, not only did John Adams and Thomas Jefferson die in the same year, on the same day, but that day was the 4th of July 1826, the 50th anniversary of the American "Declaration of Independence." More recently, at the premiere performance of Leoš Janáček's *The Makropulos Case* at the NY Met on 5 January 1996, the tenor Richard Versalle collapsed and fell ten feet to the stage (subsequently dying) after he sang the line "Too bad you can only live so long". [We thank Robert Marsh for xeroxing for us a copy of the *NY Times* obituary (7 Jan. 1996) and an article from the *New Yorker* (22 Jan. 1996), which gives a very moving, eyewitness account of Mr. Versalle's tragic death.] How many of us would be so "gullible" as to believe an account of such a astounding coincidence as either of these, if it were to appear in an ancient source?

Although no fewer than nine texts attest the conjunction of the comet with Octavian's games in 44 B.C. (Texts 1-9), and another four texts give an account of the comet without mentioning the games (Texts 10-13), this evidence has several telling features that might arouse suspicion. First is the fact that the earliest of these sources, a passage from Augustus' own *Memoirs*, was written at least two decades after the comet appeared.[4] There is no extant account of the comet in works written nearer in time to the event, apart from a few stray allusions in the Augustan poets (Texts 20-23, 31). Second, and more troubling still, closer inspection of the later sources reveals that virtually all of them are based, to one degree or another, on the account given by Augustus himself. This Augustan *interpretatio* can be readily perceived by consulting Appendix V (p. 189) where the parallels are laid out in tabular form.[5] Third, and lastly, there is a gap of nearly one hundred years between Augustus' account of the comet (ca. 24 B.C.) and our next oldest account (Text 2) in Seneca's *Natural Questions* (ca. A.D. 63).

To address the last of these points first, the gap of nearly a century between our two oldest sources need not be so troubling when we stop to consider how much of Latin literature has been lost. If Livy's account of 44 B.C. had survived, if we could still read the history of Asinius Pollio (76 B.C. - A.D. 4), another contemporary of the event, and if the works of so many other early imperial writers were still extant, we would doubtless find the comet treated by many of those authors.[6] Admittedly it is less easy to explain why the comet is not securely attested in the extant sources that predate Augustus' *Memoirs*.[7] For instance, the contemporary letters of Cicero, and his fourteen *Philippics*, which were all written after the comet appeared, make no allusion to the celestial phenomenon. Cicero's silence provides a salutary counterweight to the tradition that the comet was "seen from all lands" (Text 1) and "caught the eye of everyone" (Text 4).[8]

---

[4] The *terminus post quem* for the date of composition is furnished by Suetonius' statement (*Aug.* 85.1) that Augustus' campaign against the Cantabri in Spain (26-25 B.C.) was the last event in his life treated in the work.

[5] The dominance of Augustus' account also complicates any attempt to reconstruct the details of Caesar's will and to put into proper perspective the provision by which Octavian was posthumously adopted as Caesar's son: see Schmitthenner 31-34, 36, and passim.

[6] It was, for instance, treated by Baebius Macer (cited in Text 6), who appears to have been a contemporary of Augustus (see p. 84.61)

[7] The one and only source earlier than Augustus' *Memoirs* that *may* attest a comet in 44 is Virgil's *Georgics* in which the claim is made that numerous comets were seen after Caesar's murder (Text 31).

[8] Later (pp. 112-16) we shall discuss this potentially ominous silence and show why it need not vitiate our other sources that do attest the comet.

Equally surprising is the failure of the more contemporary Roman coinage to celebrate the comet that is credited by our later sources with instilling in the common people a belief in Caesar's divinity. Of course, we would not expect Mark Antony or the Roman Senate in 44 to embrace this notion and glorify it on Roman coins; they were opposed to elevating Caesar to the rank of a god (see p. 2.3). By the summer of 43, however, Octavian had seized control of the government, and in the following years, he and his colleagues in the triumvirate carefully selected the designs that were to appear on Roman coins so as to advance their power.[9] Yet despite this change in political control, even in those years there is no issue of Roman coinage on which a comet is depicted. Instead, Roman coins and the Augustan poets (Texts 20-24) consistently employ a star as a symbol of Caesar's new divinity: Virgil's *Caesaris astrum* ("Caesar's star") of ca. 42 B.C. (Text 20) being the earliest extant literary allusion to the comet of 44.[10] It was not until the time of the Secular Games in 17 B.C. that a comet finally appeared on some Roman coins, and then only briefly, in apparent allusion to the comet of 44.[11] In literature, it was not until some years later still, in Ovid's *Metamorphoses* (Texts 29-30) ca. A.D. 8, that a comet was at last introduced into Augustan poetry as a symbol of Caesar's apotheosis. What are we to make of the comparatively late appearance of the comet in our tradition? Could the star that Octavian is said to have added to Caesar's statue as a sign of his divinity (Texts 1, 3-6, and 9), somehow have given rise to the tradition that we find in the extant sources concerning the appearance of a comet during Octavian's games? Was that famous comet nothing more than an invention on Augustus' part?

The answer to this question must be "surely not" for at least three cogent reasons. First of all, although the bulk of our sources attesting the comet do undoubtedly follow the Augustan line, at least two sources (Texts 3 and 6) preserve traces of an anti-Augustan interpretation of the omen. We shall discuss in chapter seven the implications of this anti-Augustan view.

---

[9] For a discussion of this coinage, distinguishing Octavian's in 43 from that of the triumvirate in the years immediately following, see Alföldi (1973) 251-56.

[10] See Weinstock 377-79 for a good survey of the coins depicting a star in allusion to Caesar's divinity.

[11] The coins are those of Marcus Sanquinius (*RIC²* 337-42) and issues from two provincial mints (*RIC²* 37-38, 102), which may both have been located in Spain (Mattingly cviii-cix, 57,62; Sutherland 25-26), or possibly one was in Spain and the other in Gaul (Giard 12-13). We thank Curtis Clay for discussing these coins with us. Prior to these Roman coins, the only other Greek or Roman coin to depict a comet is a small bronze piece from the Pontic region, issued in the late 2nd cent. B.C. [?] (Imhoof-Blumer 185-87), apparently overlooked by Hazzard 422 n. 37.

Here it is sufficient to note that the mere existence of this hostile counterview clearly demonstrates that the comet was by no means a mere figment of Augustus' imagination but was instead a real historical event, which competing factions attempted to turn to their advantage in a war of propaganda. Second, it may well be asked why Augustus would choose to invent a comet, if one did not in fact appear. The traditionally baleful nature of comets should have caused Octavian to go out of his way to avoid associating a comet with himself or his games in honor of Caesar. To characterize the conjunction of a comet with Octavian's games as "the fortunate appearance of a comet" (Taylor, *Divinity* 242) misses the point by a mile. Instead, from Augustus' account of the comet in his *Memoirs*, we can only conclude that the future emperor possessed in 44 both the luck and the skill needed to turn what was potentially a very baleful omen into a powerful symbol of his adopted father's divinity.[12] This stroke of genius on Augustus' part has to be regarded as one of the most remarkable examples of "spin" control in the whole of antiquity.[13] Third and lastly, we can be certain that there was a comet in 44 B.C. because one is attested in our Chinese sources, and those accounts cannot be suspected of having fallen under the spell of the portentous Ides of March or Augustan propaganda.

The earliest extant Chinese account of the comet of 44 B.C., and the one on which all other reports from China and Korea depend, is found in the *Ch'ien Han-shu* (henceforth *HS*), *History of the Former (Western) Han Dynasty* (206 B.C.-A.D. 9).[14] This work, which was composed chiefly by Pan Ku (A.D.

---

[12] The result was that later a comet, or other bright celestial phenomenon, might sometimes be interpreted as a return of the "Julian star" and so be treated as a favorable omen (e.g., Claudian, *De quarto consulatu Honorii* 189-90). In fact, one explanation sometimes offered for the selection of 17 B.C. as the year for holding the Secular Games is that a comet may have appeared earlier in the year and been regarded as a return of Caesar's comet (Gardthausen I.2 1010). Paul Zanker (167; 171 of the German edition) goes so far as to assert that the appearance of that comet in 17 was "expected" but retracts this view in personal communication (9 July 1993), acknowledging that in antiquity it would have been impossible to predict the return of a comet. The evidence for a comet in 17 is extremely slim, however, and has recently been rejected by Gurval 284.

[13] Apparently this gift for finding a silver lining in every ominous cloud ran in the family. One is reminded of Caesar's ill-omened slip when he disembarked at the commencement of his African campaign in Dec. 47 B.C.; to dispel the superstitious fears of his soldiers who were looking on, Caesar is credited with grasping a handful of soil and crying out "I hold you, Africa" (*teneo te, inquit, Africa*, Suet. *Iul.* 59; cf. Dio 42.58.3).

[14] Later sources cited by Ho 147, such as the *Hsi Han Hui Yao* 28:4b (*Essential Records of the Western Han Dynasty*, composed by Hsü Thien-Lin in A.D. 1211) and the *Wên Hsien Thung Khao* 286:5b (*Historical Investigation of Public Affairs*, composed by Ma Tuan-Lin in A.D. 1254) merely repeat what is found in the *HS*. We thank Dr. John Major and Prof. John Rohsenow, our colleague in Linguistics, for providing us with literal

32-92) in the first century A.D., drew upon sources that preserved a selection of the records once kept by the imperial bureau of astronomy in the Chinese capital.[15] The Astronomical Observatory, or Directorate, which made and recorded the observations and reported directly to the emperor, was located in 44 in Ch'ang-an (near modern Xi'an, 34° 15'N, 108° 54'E), the capital of the Western Han dynasty.[16] The *HS* mentions the comet of 44 twice. The description in the *T'ien-wên-chih*, or "Treatise on Astronomy" (chapt. 26), tells us how the comet appeared over the course of several days; a more abbreviated account in the *Pên-chi*, or "Basic Annals" (chapter 9), simply gives the month of the sighting and region of the sky, adding nothing of substance to what we learn from the treatise.[17] Both texts state that the comet was observed in the fourth month of the fifth year of the *Ch'u-yüan* reign period, which corresponds to the dates 18 May to 16 June in the western, Julian calendar (Tung 249).

If we ask how reliable this report of a comet in 44 is likely to be, the answer is that although there is, unfortunately, no means of testing this particular record, in other years where it is sometimes possible to check the accuracy of Chinese astronomical reports, those sources prove to be remarkably reliable.[18] For example, (a) from 240 B.C. through A.D. 1378, our best accounts of 1P/Halley (1P/ indicating its status as the first comet proven to be periodic) are generally those given by the Chinese, and of the first ten returns during this period, through the year A.D. 451, records of all but one are preserved in the Chinese sources;[19] (b) in four other instances, Chinese

---

translations of these texts and for giving us the benefit of their expertise on a variety of related questions.

[15] Stephenson 238-39, with Fig. 13.3, has recently demonstrated that two main sources are likely to lie behind the astronomical records in the *HS*, one of which preserved fewer details from the imperial archives than the other. Beck 114 offers a tentative identification of one of these sources.

[16] See Needham III 189-91, quoting from the *Chou Li* (*Record of the Rites of Chou*, composed in the second century B.C.), for a description of the imperial observatory, its chief officials (astronomer, astrologer, meteorologist, timekeeper), instruments (brass armillary spheres and sighting tubes), and place of observation (a tower or platform).

[17] The text in the "Annals" is not taken into account by Ho 147 or by Zhuang-Wang 387, whose collection of texts (Chinese only, without translation) aims at being more comprehensive. It turns out that this overlooked notice in the "Annals" is the source of a much later Korean text (12th cent. A.D.) which Ho does cite and translate (see pp. 110-11).

[18] Schove xxvii remarks concerning the Chinese sources in the compilation made by Ho: "I have found the catalogue of my collaborator Dr. Peng-yoke Ho . . . very reliable, as the Far Eastern dates nearly always agree with the corrected Western dates."

[19] Ho nos. 19, 47, 61, 78, 100, 122, 156, 175, and 204 for the years 240, 87, 12 B.C., A.D. 66, 141, 218, 295, 374, and 451. Thanks to the recent discovery of Babylonian sightings of 1P/Halley in 164 B.C. (Stephenson et al.), it is now possible to be more certain that the Chinese record for 240

reports of comets happen to be confirmed by wholly independent records kept in Babylonia;[20] and finally, (c) computation has verified as accurate the vast majority of the 54 solar eclipses reported in chapter 27 of the *HS*, the *Wu-hsing-chih* ("Treatise on the Five Elements").[21]

The chances are good, therefore, that a comet was indeed sighted by the Chinese in 44 B.C., but next we might wonder whether the date of the Chinese sighting has been correctly converted to the Julian dates 18 May-16 June. Can it, or perhaps should it, be brought into harmony with the Roman sighting in late July?[22] In our view, this possibility is extremely remote for at least three reasons.

First, our understanding of the Chinese calendar in this period seems to be quite sound, and scholars are in general agreement on how Chinese dates should be converted to western, Julian notation.[23] Therefore, the fourth month

---

B.C. concerns Comet Halley: Stephenson (1990) 232, 249-50.

[20] Ho nos. 21, 26, 35, and 42 for the years 234, 157, 138, and 120 B.C. The Babylonian records (Yeomans 364-5) are, of course, extremely fragmentary, but enough survives to confirm not just the month and year of the Chinese notices but also in two of these instances even the region of the sky in which the comet appeared (in 234 and 120 B.C.). We thank Hermann Hunger for sending us, in advance of publication, a transcription of the record for 120 in translation.

[21] See Dubs, appendices to each of chapts. 1-12 of *HS*. Of these 54 solar eclipses, 39 are found to be correctly dated, and another two can be added if minor textual mistakes are assumed in the reports. This means that better than 75% of the reports are accurate, and the R. A. of the Sun, when given, is rarely off by more than 5 degrees. Of the 13 false reports, two may be in error because they were attempts to predict eclipses, and another seven fall within in a narrow 27-year period (160 to 134 B.C.) and so may reflect a temporary aberration in reporting standards. This leaves an amazingly small number of erroneous reports (4 out of 54, = 7%) which cannot be explained. We thank F. R. Stephenson for sending us by e-mail an abstract of an as yet unpublished paper written with Neasa Foley which reviews the evidence and the conclusions drawn by Dubs for the Western Han period.

[22] Brian Marsden in personal communication (22 July 1995) suggested that the Chinese and Roman sightings might be "more coincident in time than the records seem to say" since the sighting in May-June is unattested in our Greco-Roman sources and the sighting in July is unattested in our Chinese sources. In chapter 5 we address, and try to explain, this troubling silence of both bodies of evidence. In testing the hypothesis that the two sightings may possibly concern the same period, we assume that the only option is to move the Chinese sighting to July—not the Roman sighting to May-June—since (1) the date of the Roman sighting is securely fixed by the date of the festival during which the comet appeared, and (2), as stated in the text, the Roman civil calendar was in agreement with the Julian calendar in 44 B.C. Therefore, the Roman report must concern 20-30 July 44 B.C.

[23] The two standard authorities, Tung and Ch'en, are in perfect agreement on the conversion of Chinese to western dates for the period following the major reform of the Chinese calendar in 104 B.C., although prior to 104 B.C. Tung's western dates are 88 days earlier than those calculated by Ch'en. On the astronomical reforms of 104 and the accuracy of intercalation in the Chinese calendar, particularly in the Han period, see

of the fifth year of the *Ch'u-yüan* reign period of Emperor Yüan (48–33 B.C.) almost certainly corresponds to 18 May–16 June (having the Julian day numbers 1705490-519), while scholars hold that after Caesar's reform of the Roman calendar in 46 B.C. (the "final year of confusion", Macrob. *Sat.* 1.14.3), March 1st 45 B.C. of the Roman civil calendar corresponded to March 1st of the Julian calendar.[24] Accordingly, the days of the festival during which the comet appeared in 44 B.C. (20–30 July) had the Julian day numbers 1705553-563 (*SKCL*). This means that for the Chinese sighting to belong to July, rather than May-June, either our understanding of the Chinese calendar in this period is in error, or we must posit a scribal error for the number of the month.[25] Such a scribal error, causing a shift of one lunar month, the "fifth" (= 17 June–15 July in 44 B.C.) to the "fourth" (18 May–16 June), cannot be ruled out, although we can be reasonably certain that the date of the Chinese sighting was given as the "fourth month" in the source upon which the *HS* drew because that date is found two separate parts of the *HS* that were most likely composed by different authors writing some thirty years apart: in the "Basic Annals" (chapt. 9) and in the "Treatise on Astronomy" (chapt. 26).[26]

Second, leaving aside the rather remote possibility that the date of the Chinese sighting should be adjusted forward by one lunar month, from the fourth to the fifth, we find that the descriptions of what the Chinese and Romans saw are quite different. The Chinese report having seen a comet that developed a tail of 10 degrees or more, whereas the object seen by the Romans is described by one writer, who was most likely a contemporary observer, as being "a very large star . . . that was *surrounded* with rays, like streamers on a garland" (Text 6). In other words, the Roman sighting concerned a comet that apparently had no discernible tail but shot out bright rays in all directions (so described in Text 13), as comets sometimes do. The celestial body observed by the Romans in July must have lacked a tail because otherwise it is impossible to explain how disagreement could have arisen among contemporary observers over whether the bright light in the sky was indeed a comet or a new "star" (Texts 3 and 6).

Cullen's 1993 article.

[24] See Tung 249 for the Chinese calendar. In the West, 1 Jan. 45 B.C. of the civil calendar corresponded to either 1 Jan. of the Julian calendar (LeVerrier and Soltau), or 2 Jan. (Holzapfel-Groebe, Unger, and Brind'Amour), depending upon whether the year is reckoned as *bisextilis* (*SKCL*).

[25] We thank Christopher Cullen for discussing these two possibilities with us in personal communication (11 Feb. 1994).

[26] For the separate authorship and dates, see Beck 12, 112-14. We thank Prof. Jack Dull of the University of Washington for discussing with us the composition of the *HS* and for consulting the ancient commentaries on the relevant passages to verify that they have nothing of importance to add to the texts themselves on our topic.

Third, it is difficult, if not impossible, to make the location of the comet that is given by our Chinese sources fit a date in July, as opposed to May-June. According to those sources, the comet was seen in the NW, in *Shen* (a region of the sky in the vicinity of Orion, near the ecliptic, bounded by right ascensions of 3h 49m and 4h 21m in 44 B.C.), and the comet's tail pointed NE. This means that the comet must have been north and east of the Sun, since a comet's tail points away from the Sun. However, by 20-30 July the Sun would have been nearly 4 hours east of *Shen*, and so the comet could not have been both NE of the Sun and in the vicinity of the constellation Orion at the time when the Romans report having seen the comet (approx. 5:00-6:15 P.M.). By that time of day, *Shen* would have been well below the western horizon. On the other hand, in late May the Sun would have been located within the division *Shen* (see Figure 1), and so if the comet was observed in the NW, NE of the Sun, it would indeed have been in *Shen*.

Granted, then, that the Chinese and the Roman sightings did in fact take place at different times of the year, nearly two months apart, what are the chances that the comet seen from China in the late spring is identical with the one seen from Rome in the summer? Before we try to answer this question, it is fair to say that the Romans themselves apparently regarded their comet as unique and distinct from any other that may (or may not) have been observed by them earlier in the year. We draw this conclusion from the fact that Caesar's comet is said to have been visible for only seven days (Texts 1, 9, 10, and 12), taken with the assertion of Pliny the Elder writing in the first century A.D. that "the shortest span on record during which a comet had been observed was seven days" (*brevissimum* . . . .*VII dierum adnotatum est, NH* 2.90). Of course the understanding of comets in antiquity was so limited that the Romans could easily have been deceived into thinking that what they saw in July was unrelated to any object seen earlier in the year. They could have been misled particularly if a gap separated the observations and if the comet appeared suddenly in July, in a region of the sky where no object had been observed in the immediately preceding period. Comets typically disappear for a time when they pass perihelion, and when they become visible again to the naked eye they can change over from evening to morning, or morning to evening objects, thereby confusing observers into thinking that they are seeing two discrete and unrelated objects. [27] Therefore,

---

[27] As recently as the seventeenth century, Sir Isaac Newton made several false starts in developing his theory of cometary dynamics based upon his observations of the comet in 1680 because at first he interpreted the appearance of that comet in the morning during November, and then later in the evening during December after it had passed perihelion, as two different objects (Yeomans 95, 99).

we cannot conclude that Caesar's comet was necessarily unrelated to the comet sighted earlier in the year by the Chinese simply because our Greco-Roman sources tell us that Caesar's comet was visible for only seven days.

When we turn to modern scholarship, we find that sometimes the Chinese and Roman sightings are treated (without argument) as belonging to the same comet,[28] and sometimes they are assigned to two separate comets. Those who adopt the "two-comet" interpretation do so, it appears, chiefly under the lingering influence of a now discredited theory that Caesar's comet was periodic, having a period of 575 years. The author of that theory was none other than Sir Edmund Halley, who identified the comet of 44 B.C. with the comets of A.D. 531, 1106, and 1680 simply on the basis of their reported physical characteristics and the intervals between them.[29] Halley's views were widely accepted for a time and even led William Whiston (1667-1752), Newton's successor as Lucasian professor of mathematics at Cambridge, to conclude that Caesar's comet was identical with a comet that had supposedly made a near approach to Earth on 28 November 2349 B.C. [!] and caused the Biblical Flood of Noah. In the first two editions of his book *A New Theory of the Earth* (1696; 1708), Whiston had not attempted to identify his comet of 2349 with one in historical times, but in the third (1722) and subsequent editions (ed. 1725, pp. 185-97) Whiston confidently stated that the diluvial comet and the comet of A.D. 1680 were identical—a noble pedigree, indeed, for the comet of 44 B.C.[30] Halley's theory concerning the 575-year period of

---

[28] For instance, Levy 17 describes the comet seen two months after the Ides of March, "with a tail perhaps 12 degrees long" (clearly the date and description of the comet seen by the Chinese), as "one of the best known from ancient times" because of the role it played in Roman history after the murder of Caesar. So too Yeomans 367 and K. Fitzler-O. Seeck, *RE* 10.1 (1917), 218 lump the Chinese reports with the Roman as if they clearly concerned the same comet.

[29] Halley's conclusions concerning the periodic nature of Comet Caesar were first presented in the 1715 English edition of his *Synopsis* 901-3; they were not included in either of the 1705 Latin editions (Oxford and *Philosophical Transactions of the Royal Society of London* 24, pp. 1882-99) or in the 1705 English edition (London).

[30] According to Yeomans 164, Whiston learned of Halley's conclusions from the second edition of Newton's *Principia* (1713), but while Whiston (191) does cite Newton's 2nd ed. (p. 465) for the view that the comet of 1680 had a period in excess of 500 years, he specifically cites and quotes pp. 901-903 of Halley's *Synopsis* as printed in Gregory's *Astronomy* (1715) for his facts concerning the supposedly periodic nature of the comet of 1680. It was not until the 3rd ed. of Newton (1726) that Halley's conclusions were included in the *Principia* (vol. 2 pp. 351-2 of the English ed. by Motte, 1729). (We thank Donald Yeomans for discussing the subject of Whiston's source with us by e-mail.) Whiston (191) was intrigued by the facts that the interval of 4028 years between his hypothetical comet of 2349 B.C. and the comet of A.D. 1680, when divided by 7, yielded a

the comet of 44 also made its way into the Diderot's *Encyclopédie*, and from there it was taken up by Edward Gibbon in his *Decline and Fall*. Writing on the comet of A.D. 531 under Justinian, Gibbon asserted in chapter 43 ad fin. (IV 289-91) that up to seven past visits of Caesar's comet could most probably be distinguished (in 1767, 1193, 618, 44 B.C. and A.D. 531, 1106, 1680), with a return yet to come "in the year two thousand three hundred and fifty-five [*sic*]."[31]

Although further study of the comet of 1680 proved conclusively that Halley had grossly underestimated its period and so it could not be identical with the comet of 44 B.C.,[32] many of Halley's views had been embraced in the meantime by the French scholar Alexandre Pingré (1711-1796) in his *Cométographie* of 1783-84, a pioneering work that took into account reports of comets both in European and East Asian sources. Pingré (I 278-9) assigned the Chinese sighting in May-June 44 B.C. to a separate comet because otherwise the comet seen by the Romans (in September, as Pingré believed, in keeping with Halley's interpretation of the evidence) could not have been identical with the comet of 1680.[33] Subsequent lists and catalogues of comets commonly distinguish two separate comets in 44 (one seen by the Chinese in the spring and another seen by the Romans in the autumn),[34] yet

---

period of approx. 575.5 years for each return and (p. 195) that both comets had supposedly made a very close approach to Earth—Halley had erroneously calculated that the comet of 1680 had passed the Earth within the range of the Moon.

[31] This must be a slip on Gibbon's part for A.D. 2255 (1680 + 575 yrs.) since he accepted the notion that the comet had a period of 575 years, and as our colleague Alexander MacGregor has pointed out to us, the date is spelled out in the standard editions of Gibbon; if the date had been expressed in numerals, one might suspect a typographical error. The source of Gibbon's information, as he tells us (289 n. 76), was the article "Comète" by M. d'Alembert "in the French Encyclopédie [edited by d'Alembert and Diderot]". We thank Peter Conroy, our colleague in French, for discussing with us the significant role played by d'Alembert as editor of the early volumes, to which he contributed many of the articles on scientific subjects.

[32] The German astronomer Johann Encke (1791-1865) determined that the probable period of the comet of 1680 is in the range of 8800 years, with a minimum not much less than 2000 years (*Zeits. f. Astr.* 6.157).

[33] In personal communication (13 May 1996) Brian Marsden points out that Halley was not aware of the Chinese sighting when he tried to establish a connection between the Roman comet and the comet of A.D. 1680. What Halley tried to show would not work for the Chinese sighting, and since Pingré accepted Halley's conclusion concerning the connection between the comets of 44 B.C. and A.D. 1680, Pingré was forced to dismiss the possibility that the comet seen by the Chinese could have been an earlier sighting of the comet observed by the Romans.

[34] E.g., Chambers 556, comets nos. 69 & 70, Baldet 16, nos. 122 & 123, and most recently Hasegawa 65, nos. 122 & 123. All three authors explicitly cite Pingré, whose work, although more than two centuries old, continues to be held in high regard by astronomers (see p. 3.9).

the sole reason for doing so appears to be the reliance of these later works on Pingré, who in turn was trying to make the evidence fit the now discredited notion that the comet of A.D. 1680 was a return of the comet of 44 B.C. It is high time that we free ourselves from this tralatitious error. Let us, without any preconceived notions, consider whether it is more probable that the Chinese and Roman sightings concerned the same comet or two separate ones.

The July sighting in 44 was of an extremely bright, star-like object, surrounded by a slight radial coma, that maintained its high luminosity for at most seven days (see p. 85). This is almost certainly the signature of an "anomalous outburst" (Richter [1949]).[35] Despite the name, anomalous outbursts are not particularly rare. Vsekhsvyatskii (cited by Hughes 412) estimates that out of 79 recent comets, 59 have had outbursts. Even Comet 1P/Halley underwent a 9 magnitude increase in luminosity, on 15 February 1991, when it was some 14 A.U. out from the Sun (Meech).

Although cometary outbursts do occur with some frequency, there is no record of one that was daylight visible. Nevertheless, it is clear that many outbursts that have been observed would have been daylight visible, if the comets had been closer to the Earth when the brightening took place. For instance, Comet P/Schwassman-Wachmann I travels in a nearly circular orbit between 5 and 8 A.U. On 23 February 1963, it had an outburst that would have given it a magnitude of -3.5 if its position had been 1 A.U. from both the Sun and the Earth (Hughes). A little closer to the Earth, 0.8 A.U., and its magnitude would have been a daylight visible -4.[36]

The question then becomes, do we have in 44 B.C. one comet that was seen twice, approximately two months apart, the second time while undergoing an outburst,[37] or were there two comets, one observed during perihelion in May, and the other in July while undergoing an outburst? Of

---

[35] The evidence could also be interpreted as concerning a nova or supernova, rather than a comet, but this possibility is rather remote: see Appendix VI.

[36] The magnitude of a nonself-luminous object, at solar distance r and Earth distance Δ, is given by $m = m_0 + 5\log(r) + 5\log(\Delta)$, where m_0 is its magnitude at $r = \Delta = 1$ A.U.

[37] In personal communication (10 Oct. 1996), Gary Kronk has kindly drawn our attention to what may be an analogous gap of well over two months between reported observations of the same comet (Ho no. 191): on 24 June A.D. 418 the Chinese saw a "sparkling comet" (*hsing-po*: on this term, see p. 110.54) and on Sept. 15 a "comet having a tail" (*hui-hsing*). Kronk concludes that these sightings most likely concern the same comet from the statement of the Greek writer Philostorgius (*Hist. eccl.* 12.8), a contemporary observer, who saw a cone-shaped comet during a solar eclipse on July 19 and writes that the comet "at length disappeared, after it had continued its course for more than four months," adding that "it began about midsummer and continued until nearly the end of autumn".

course without a known orbit we cannot prove that the two sightings in 44 B.C. were of the same comet; however, we can demonstrate that it is somewhat more probable that it was the same comet rather than two different ones. Hasegawa's compilation, combined with the supplement by Jansen, lists 236 naked-eye comets seen from the northern hemisphere during the final 270 years covered by his catalogue, the period for which our records are the fullest (1701-1970). This is a rate of r = 236/270 = 0.87 comets per year. Now, unless there is some strange mechanism scheduling comets, comet sightings should obey Poisson statistics.[38] These statistics apply to any process that occurs at a definite rate, but is otherwise completely random. Knowing the rate of occurrence, we can use the Poisson theory to estimate the probability of occurrence of n comets [$P(n)$] within a time interval of any given length. Given this probability, and knowing the number of all events, we can estimate the number of times a set of n comets would be seen within such a time interval. For instance, from the above rate r = 0.87 comets per year, Poisson statistics predict that during a period of 270 years there should be seen 32.9 ± 5.7 pairs of comets whose maximum brightness occurs within two months of each other.[39] From Hasegawa and the supplement by Jansen, the number of such comet pairs during the 270 years through and including 1970 is actually 27, in good agreement with the Poisson prediction.

Now, the Romans saw a comet around 23 July. About two months previously the Chinese saw a comet. Given the Roman sighting in July, the probability that the Chinese and Roman comets are the same is the probability that no other comet should have been seen during the two-month interval, namely (see n. 39),

P same = P(0) = 0.86.

---

[38] Reif 41-42. According to Poisson statistics, if r is the rate of comet appearances and Dt a time interval, the mean number of comets in Dt is $<N>$ = rDt, and the probability of exactly n comets being seen during Dt is $P(n)$ = $<N>^n$ exp($-<N>$)/n! Here, the mean number of comets seen in a two month interval is $<N>$ = **0.87 comet/year x 2/12 yr = 0.15**.

[39] If one comet appears, the probability that at least one other comet will appear in a prescribed time interval, P_{more}, is P_{more} = 1 - $P(0)$. $P(0)$ is the probability that no other comet will appear in the time interval. Given N_p successive comet pairs, the expected number of these pairs that appear within the time interval Dt, $<N_{p,Dt}>$, is then $P_{more}N_p$. This will be uncertain by an amount equal to its square root. The probability of no other comets being seen during a two month interval is $P(0)$ = exp($-<N>$) = exp(-0.15) = 0.86. Thus P_{more} = 1 - $P(0)$ = 1 - **0.86** = **0.14.** With 236 comets seen, the number of successive pairs is N_p = 235, then $<N_{p,Dt}>$ = $P_{more}N_p$ = 0.14 x 235 = 32.9, and $\sqrt{32.9}$ = 5.7.

The probability that the comets are different equals the probability that at least one other comet should show up during that time interval:

P diff. = 1 - P(0) = 0.14.

It is, therefore, a little more than six times more probable that the two appearances were of the same comet than that they were of different comets (0.86 ÷ 0.14).

For the purposes of this study we shall henceforth assume that we are dealing with a single comet, and we shall treat the Chinese report as giving us a second point of reference to compare with what we are told about the comet when it was seen by the Romans in late July. Of course, two positions are not sufficient to determine a comet's orbit.[40] However, from the relatively detailed account preserved in our Chinese sources, we can draw certain inferences about the comet's approximate date of perihelion and perihelion distance. Starting from these assumptions, we can fit the two positions indicated by the sightings in May-June (from Ch'ang-an) and late July (from Rome) with a range of possible orbits, and hence arrive at a likely approximation to the orbit of Caesar's comet.

For the convenience of readers who do not control the technical side of astronomy, each chapter division that presumes a knowledge of the mathematics of astronomy is provided with a less technical abstract. In these précis we try to present in layman's terms both the major pieces of evidence and the conclusions that we draw from this evidence so that the general reader can follow the lines of our argument about the comet's likely orbit. We also refer the reader directly to the relevant figures and graphs that portray the regions of the sky in which the comet is reported to have been observed, that offer reconstructions of its likely orbit, and that provide illustrations of its estimated brightness and times of rising and setting as viewed from the Chinese capital and from Rome.

---

[40] See McCuskey 70-91. A few tantalizing cometary records have been brought to light in the cuneiform tablets from Babylonia, two volumes of which have been edited and published by Abraham Sachs and Hermann Hunger under the title *Astronomical Diaries*, but Prof. Hunger confirms by letter (6 July 1993) that the records that have so far been uncovered in the Middle East break off with the year 47 B.C. We cannot, therefore, hope to learn more about the comet of 44 from that source unless some day more tablets are recovered (a possibility that Prof. Hunger does not rule out, if conditions in Iraq permit further scholarly explorations). Likewise we have verified that the astronomical records from Korea, both those published by Tamura 128-37 and those being assembled by Il-Seong Nha and a team of scholars, offer no additional information about the comet of 44 beyond what our Chinese and Greco-Roman sources tell us. We thank our colleague Kyoko Inoue for translating the relevant record in Tamura, and we thank Prof. Nha for sharing his findings with us by letter (18 Jan. 1994).

THE OBSERVATIONS OF CAESAR'S COMET

The Chinese Sighting

Abstract

The Chinese report describes an observation that was made over the course of several days during the period of 18 May to 16 June. We can tell from the statement that the comet was seen in the NW, from its location in the sky (N of the constellation Orion), and from its relative closeness to the Sun (Figure 1), that this was an evening sighting and that the comet was near its perihelion (estimated dates given in Table 1). We infer from its reported position relative to the Sun, combined with the fact that the comet seems to have been visible for only a few days, that it probably developed its tail (ca. 8° when first reported) as it approached the Sun from the south and remained at first deep in the sunset skyglow, emerging for only the few days alluded to in the Chinese report. It will then have been invisible for a few days because it was too close to the Sun at sunrise and sunset to be seen, re-emerging in early June as a moderately bright-to-faint object that rose in the NE at an ever increasing interval in advance of the Sun (Figure 2).

From the geometry of the inferred position relative to the Sun in May-June (as shown in Figure 2), we can estimate that its distance from the Sun at perihelion was at most 1/4 A.U. (1 A.U. [astronomical unit] being equal to the mean radius of the Earth's orbit, approximately 93 million miles or 150 million kilometers). The small size of this perihelion distance makes it very likely that its orbit was nearly parabolic and, therefore, that the comet of 44 was either a long-period comet (over 200 years) or nonperiodic.

As previously indicated (p. 66), two accounts of the comet of 44 B.C. are preserved in the *HS*. A brief mention in the "Basic Annals" (chapt. 9) tells us only that a comet was seen "in the fourth month" in a region of the sky known as the *Shen*. The fuller description in the "Treatise on Astronomy" (chapt. 26) agrees with the statement in the "Annals" and tells us quite a bit more about the comet. We offer a literal translation of the passage from the "Treatise", enclosing in pointed brackets words and endings that are required in English but not expressed in Chinese characters. (Classical Chinese texts tend to be "telegraphic" in nature.) The reference in square brackets to the *Ch'u-yüan* reign period (48-44 B.C.) is supplied from earlier in the chapter.

"<In the> fourth month <of the> fifth year [of the *Ch'u-yüan* reign period], <a> broom star (*hui-hsing*) appear<ed in the> NW, red<dish->yellow <in> color

<and> eight *ch'ih* long. Several days later <it was> more <than one> *chang* long, 'point<ing> NE, locate<d in the> division <of the> *Shen*." (*HS* 26:31b)[41]

As we have already remarked, the fourth month in the fifth year of the *Ch'u-yüan* reign period corresponded to 18 May-16 June 44 B.C. in the Julian calendar, and there is little likelihood that this report concerns the same period as our Greco-Roman sources which describe how the comet appeared in July (pp. 67-69). A "broom star" (*hui-hsing*) was one of the names for a comet (*hui*, lit. "broom"), regarded by the Chinese as one of several ominous stars. Ho (136) gives the following account, translated from the *Chin-shu* 14:4a (*Official history of the Chin dynasty*, completed in A.D. 635): "Its body is a sort of star, while its tail resembles a broom. Small comets (*hui*) measure a few inches (*ts'un*) in length, but the larger ones may extend across the entire heavens."

The *ch'ih*, a unit of linear measurement (= approx. 9 English inches in the Han period), when used in astronomical records has in the past been thought to be equivalent to 1.5 degrees of arc (Kiang 1972). The latest research, however, concludes that each *ch'ih* is to be regarded as more nearly equivalent to one degree,[42] and so our comet's tail will have measured approximately eight degrees at first, and a few days later approximately ten degrees (one *chang* = 10 *ch'ih*). This would be a relatively short tail as comets go, ten degrees being equivalent to roughly twice the distance between the "pointer stars" in the Big Dipper. By contrast, one of the most

---

[41] Since scholars still sometimes consult the collection of Chinese cometary records compiled and translated by John Williams (1871), we should state that Williams' account of the comet of 44 B.C. (p. 9) is unreliable (errors indicated by italics): "a comet appeared in the *north-east*; its colour was reddish yellow. It was 8 cubits in length. A few days after, its length was about 10 cubits. It was then *in the north-east, pointing towards* the S(tellar) D(ivision) Tsan [an alternate Romanization of the Chinese character *Shen*]. *After about two months(?)* [*sic*] *it turned again to the west*." We have consulted the Chinese text that Williams translates, the *Wen Hsien Thung Khao* [*WHTK*] of the 13th cent. A.D. (see Dubs ap. Rogers 239 n. 6 and Ho 127), and find that Williams has misrepresented what the text actually says: the text of the *WHTK* is, in fact, identical to the text of the *HS* translated above. It states that the comet "appeared in the NW", that it "pointed NE" and was "located in the division *Shen*". The text goes on to say that "after two more years the western Chiang revolted", which Williams mistook for a reference to the comet, changing "two years" to "two months(?)" presumably in an attempt to make the text fit his preconceived notion.

[42] Wu-Liu, *Publ. Shanxi Observatory* (1990). We thank Dr. Huang Yi Long of the National Tsinghua Univ. in Hsinchu, Taiwan, for a summary of the findings of this publication in communication by e-mail (1 Feb. 1994).

spectacular comets in this century, concerning which anecdotes still abound,[43] is Comet Halley during its return in 1910, when it developed a tail of 100 degrees (Hasegawa 90). Such a comet as Halley's could not fail to be noticed even by disinterested observers, but as we shall see (pp. 95-99), the comet of 44 apparently made little or no impression upon observers in the Mediterranean at the time of the Chinese sighting in May-June.

Finally, *Shen* is the name of the 21st "lunar lodge" or "lunar mansion", one of the divisions of the heavens. Needham (231-32) gives the following description of these celestial divisions: "During the 1st millennium B.C. the Chinese built up a complete system of equatorial divisions (the celebrated lunar mansions), defined by the points at which these hour-circles (radiating from the poles) transected the equator—these were the *hsiu*. One has to think of them as segments of the celestial sphere (like segments of an orange) bounded by hour-circles and named from the constellations which provided determinative stars, i.e., stars lying upon these hour-circles, and from which the number of degrees in each *hsiu* could be counted. . . . Now once having established the boundaries of the *hsiu*—and for this the declinations of the *hsiu*-asterisms and the determinative stars, whether nearer or farther from the equator, did not matter at all—the Chinese were in a position to know their exact locations, even when invisible below the horizon, simply by observing the meridian passages of the circumpolars keyed to them."[44]

The marker star in the constellation *Shen* ("White Tiger"), after which the 21st "lunar lodge" is named, is one of a group of stars that today make up Orion's belt. The next "lunar lodge" to the east (the 22nd) is *Tung-Ching*, named for a constellation that is roughly equivalent to Gemini. Needham (234-37) in his Table 24 gives the marker star for the division of the *Shen* as ζ *Orionis*, and that for the division of the *Tung-Ching* as μ *Geminorum*,[45] but

---

[43] For example, one of the APA referees remarked in his report on our MS that "my great uncle and his neighbors left buckets of water on their roofs to keep them from burning up [when Comet Halley appeared in 1910]!" Such measures are unlikely to have been deemed necessary in response to the comet of 44 which, in our view, never developed a spectacular tail of the type seen in 1910.

[44] Needham may overstate the latter point since only a few *hsiu* have been shown to have been linked with circumpolars: Kiang (1984)—we thank F.R. Stephenson for the reference. Presumably, however, stars closer to the *hsiu* would have permitted skilled Chinese astronomers to *estimate* boundaries when the marker stars themselves were below the horizon (see p. 78).

[45] The marker star of each *hsiu* ("lunar lodge") is the star "lying on the hour-circle with which it [the lunar lodge] begins", Needham 234. We thank Professor Ho Peng Yoke, Director of the Needham Research Institute, for commenting by letter (29 Oct. 1993) on some of our earlier conclusions concerning the boundaries of the division of the

as Needham later remarks (251), prior to an innovation during the Ming dynasty (probably fairly late, circa A.D. 1628[46]) the marker star for *Shen* had been δ *Orionis*. Therefore, for the period in which we are interested, the division of the *Shen* would be that part of the celestial sphere having its western boundary defined by δ *Orionis* and its eastern by μ *Geminorum*. In 44, this would have been between 57.2 and 65.2 degrees of right ascension, an interval of 8 degrees.[47]

On the 30th of May, near the midpoint of the given time interval (18 May-16 June), the Sun would have been at 63 degrees of right ascension, well within the division of the *Shen*. The ecliptic runs rather close to μ *Geminorum*, and on that date μ *Geminorum* would have been within 5° of the Sun and therefore invisible. Moreover δ *Orionis* would have been below the horizon at sunset, and thus the *Shen*'s marker stars could not have been visible and its position would have to have been estimated from the surrounding stars (see n. 44). It is doubtful whether it could have been estimated to better than ±5°.[48]

The comet's tail is said to point to the NE. The tail has to point away from the Sun, so the comet must be seen to the NE of the Sun. This rules out the comet actually being in the asterism of *Shen* since Orion was south of the Sun,[49] as shown in Figure 1, which depicts the northwestern sky as seen from Ch'ang-an at 45 minutes after sunset, on 30 May 44.

If the comet was seen in the NW and was NE of the Sun, this has to be an evening sighting, shortly after sunset, with the Sun below the horizon in the west. The term "NW" covers a lot of territory. We would take this to be somewhere between WNW and NNW, but the Chinese in 44 may easily have used a different convention. For a latitude of 34° N, an object on the horizon within this range of azimuths would have a range of declinations from 18.5° to 50°. A comet on the horizon exactly NW of the observer would have a declination of 35.9°.[50]

Shen and on other matters of a technical nature.

[46] F.R. Stephenson, personal communication (5 Jan. 1995).

[47] See for a recent discussion, Maeyama and Stephenson (1994).

[48] F.R. Stephenson, personal communication (5 Jan. 1995).

[49] Modern cometary catalogues typically state that in May-June the comet of 44 was seen in the region of Orion: Baldet 16, Hasegawa 65, Yeomans 367.

[50] From the Chinese capital, the view to the NW should have been relatively unobstructed by the terrain of the Wei River valley which is quite flat (nothing over ca. 100 m.) in that direction for a distance of 30 to 35 km. to the horizon: see U.S. Army Map Service, "China Proper NW 1:250,000 (A. M. S. L532)" 1945. sheet I 49G "Chien Hsien". We thank the UIC map librarian Marsha Selmer for helping us investigate that region of China, and we thank Richard Bates for discussing with us, by correspondence, the factor that terrain may have played in the Chinese sighting.

We conclude, therefore, that the comet when observed in China was somewhere within that part of the celestial sphere bounded by right ascensions 57.2° and 65.2°, and declinations 18.5° to 50° (= WNW to NNW at 34° N latitude). This is an area of the sky between Gemini and Auriga. It is shown shaded in Figure 1.

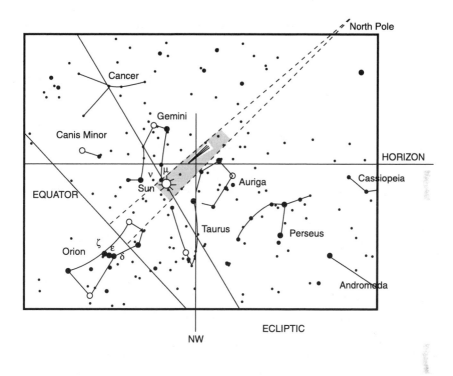

Figure 1: The view to the northwest, as seen from 34° N (the latitude of the Chinese capital rounded to the nearest whole degree), at 45 minutes after sunset, on 30 May 44 B.C. The northern half of the division of the Shen is delineated by the dashed lines. The likely area for the comet's position is shown shaded. This is a central projection onto the plane normal to the NW azimuthal direction. It is what would be seen by a camera with a 100° wide angle lens aimed to the NW. There will be some distortion towards the edge of the picture. Magnitude ≤ 1, ○; magnitude 2, ●; 3, •; magnitude ≥ 4, ·.

The Chinese report has the comet appearing for only a few days at some point during 18 May to 16 June. There are a number of possible ways to interpret this evidence. (a) The comet was actually visible to the Chinese for a much longer period but only a partial report has come down to us. (b) The comet would have been visible for a longer period, but the weather was only clear for a few days. (c) The comet was too dim to be seen for much of its approach to the Sun but experienced a sudden outburst near perihelion, which lasted for only a few days. (d) The comet emerged for only a few days from the sunset skyglow.

Interpretation (a) cannot be ruled out. In fact, this particular comet is known to have inspired the emperor to issue an edict in which he apologized for the deficiencies of his rule.

> "Since We are inadequate [to Our position], the ranking [of persons] in their positions is not carefully scrutinized, and many offices have long been unoccupied and have not been filled with the [proper] persons, so that the great multitude has lost its hope [of good rulers. This situation] has affected August Heaven above, so that the Yin and Yang have produced grievous vicissitudes, hence [Our] fault has spread to the many common people. We are greatly dismayed at [this situation]. . . ."[51]   (HS 9:5b)

There follows a list of reductions in expenditures, regulation of allotments to members of the imperial household, and reduction in punishments, all ostensibly aimed at ameliorating suffering caused by plague, famine, and inclement weather.[52] The report could very well break off on the day of the apology.

Likewise, (b) and (c) are quite possible, but give us no further information on the comet's position.

In the following we shall adopt interpretation (d). That is, we assume that the path of the comet is close to the Sun during May-June, as shown in Figure 2. This figure is a schematic plot of the sky near the Sun in terms of right ascension and declination relative to the Sun. The western horizon at sunset is shown and anything to the right of that line sets before the Sun. The

---

[51] Tr. Dubs II 313-14.

[52] The Confucians played a prominent role as ministers in the government of Emperor Yüan (48-33 B.C.), and during the first few years of his rule there seems to have been a succession of droughts. It is a Confucian notion that Heaven sends a visitation (tsai) to warn when the government needs correcting. If the warning is ignored, a prodigy (yi), such as a comet, solar eclipse, drought, or earthquake, is sent to frighten the offending party. See Dubs II 286-87. The announcement of prodigies may have served as veiled criticism of the emperor: see de Bary 186-87. The imperial edict of contrition following such announcements is perhaps also to be viewed as a stock device for introducing and justifying new policies: see Wolfram 51-52.

eastern horizon at sunrise is also shown and anything to the left of that line rises after the Sun. An object in the region of the sky to the right of the western horizon and to the left of the eastern horizon both sets before, and rises after, the Sun. An object in that region, unless it is extremely bright, cannot be seen since it is above the horizon only when the Sun is also in the sky. We call that region the "cone of invisibility." The exact boundaries of that region depend on the observer's latitude and also on the day. The figure is calculated for the latitude of the Chinese sighting, 34° N, near 30 May.

The dashed lines of "first visibility" sketched in Figure 2 represent the limits placed on the visibility of a moderately bright object by the twilight skyglow at sunrise and sunset. The exact boundaries of those curves depend on the brightness of the object, the condition of the atmosphere and the inclination of the ecliptic.[53] The curves shown in Figure 2 are only schematic. Near the Sun, the curves have been drawn 15 degrees from the Sun, a limit suggested by taking the average of astronomical and nautical twilight, which occurs when the center of the Sun's disk is 18 or 12 degrees respectively below the horizon.[54] The lines then approach the horizon as one moves away from the Sun since the sky is darker there. An object exactly on the curve has its first visibility only when it is on the horizon, with the Sun below the horizon. An object between the horizon and its line of first visibility—wherever that curve lies, either as shown in Figure 2, or closer to or farther from the horizon, depending upon the brightness of the object—cannot be seen.

---

[53] Claudius Ptolemy, *Almagest* 8.6.

[54] *The Astronomical Almanac, for the Year 1992*, Washington, U.S. Government Printing Office, M13.

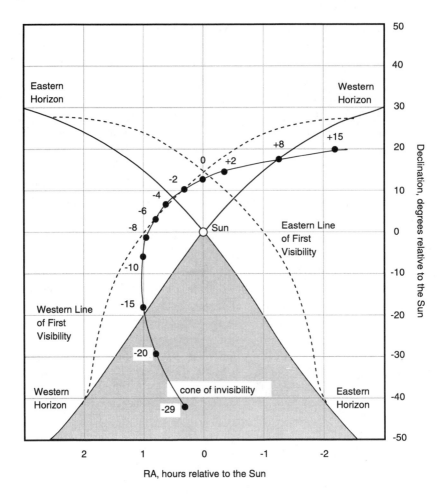

Figure 2: A hypothetical orbit for Comet Caesar, plotted in right ascension and declination relative to the Sun. The western horizon at sunset and the eastern horizon at sunrise are shown. Anything to the right of the western horizon line sets before the Sun. Anything to the left of the eastern horizon line rises after the Sun. The lines of first visibility are drawn very schematically. The orbit of our hypothetical comet is marked by the number of days relative to the comet's transit through the Sun's meridian. This orbit is retrograde and has a perihelion of 0.224 A.U.

Figure 2 shows one of our hypothetical comets (perihelion = 0.224 A.U.).[55] We have labeled the days of the orbit relative to the time of passage through the Sun's meridian. The comet develops its tail over a few weeks in the "cone of invisibility", to the south of the Sun. On day -15 it emerges from the cone of invisibility into the evening sky, but at that time it is too deep in the sunset skyglow to be seen. It sets approximately 10 minutes after the Sun. From day -6 to -2 it runs along the line of first visibility. If the comet is more than moderately bright, its line of first visibility will be closer to the horizon, and the comet will be visible for some time prior to its setting, which occurs at approximately one hour after sunset on days -6 to -2. On day 0 it is on the Sun's meridian, and just inside the line of first visibility in both the morning and evening, and is equally close to the horizon at both sunset and sunrise. From that point on, it moves deeper and deeper into the sunset skyglow. By day +8 it sets with the Sun, and from that time onwards it will set before the Sun at an ever increasing interval, as shown by its position to the right of the western horizon line in Figure 2.

We assume that the Chinese report breaks off when the comet became invisible during the evening. Depending on the object's brightness, there may have been a few days, perhaps from about day -1 to +1 according to Figure 2, when it was invisible in both evening and morning. Although this very schematic representation of the skyglow does not permit us to time precisely the comet's visibility, it does seem plausible that day 0 could not have been more than one or two days from the time of evening invisibility. If so, then the Chinese report breaks off within a day or two of the passage of the comet through the Sun's meridian, and we can narrow further the date of the sighting from China by taking the following circumstances into consideration. Since the Shen has a width of 8 degrees (57.2° to 65.2° R.A.), the Sun will have been in the Shen for about 8 days (ca. 24 May-1 June 44 B.C.). Allowing for a few days uncertainty (±3) in the position of the comet relative to the Sun's meridian, this leaves a range of about 14 days, and so the best estimate of the Chinese sighting is 28 May (when the Sun was near the midpoint of the *Shen*) ± 7 days. In arriving at our hypothetical orbit, we have assumed that the comet was on the Sun's meridian on 30 May (near the midpoint of the Chinese fourth month, and just a day or two after the Sun crossed the midpoint of the *Shen*).

The comet's period of visibility includes the few days spent grazing the skyglow's outer edge. From the geometry we can estimate a perihelion

---

[55] For the calculation of the orbital parameters, see pp. 124-29.

distance of at most 0.26 A.U.[56] We can also make a dynamical estimate, based on the period of visibility, that gives a perihelion distance of about the same size.[57] Our estimate of the perihelion distance may be used, in turn, to draw a further conclusion about the probable nature of the comet's orbit. Given the small perihelion distance, it seems likely that the orbit of Comet Caesar was nearly parabolic since of the 855 comets listed in the Marsden's *Catalogue of Cometary Orbits,* only two periodic comets have a perihelion this small,[58] but 81 of the nearly parabolic comets do. The latter are comets that have very long periods, over 200 years, and it is likely that Comet Caesar falls into this category.

The Roman Sighting

Our Greco-Roman sources may be combined to give a reasonably full account of how, when, and where the comet appeared in July. The considerable degree of agreement among these sources is best explained by the fact that most of them drew upon a tradition that was grounded in Augustus' account of the comet in his *Memoirs, De vita sua.*[59] The reported facts are these:

1. The comet became visible "at about the 11th hour" (= approx. 5-6:15 P.M. local solar time[60]): *circa undecimam horam* , Text 1, cf. 2, 4, 9. The hour was apparently given less accurately as the 8th (= approx. 1:15-2:30 P.M.) by Baebius Macer (cited in Text 6), who seems to have been a contemporary observer.[61] (See below, p. 130, for a possible explanation of this

---

[56] If $\Delta\theta$ denotes the angle seen between the comet and the Sun, and α the angle between the comet-Sun line and the Earth-Sun line, its distance from the Sun in A.U. is d $= \sin(\Delta\theta)/\sin(\alpha - \Delta\theta)$. This is about 0.26 A.U. for $\Delta\theta \approx 15°$ and $\alpha \approx 90°$.

[57] If the comet moves into and out of visibility within a few days, then it must change its true anomaly v by a substantial amount, say from 0 to 45° in Δt days. Its perihelion distance q $=\{0.0365\ \Delta t/[\tan^3(v/2) + 3\tan(v/2)]\}^{2/3}$. This is 0.27 A.U. if we take Δt to be five days. See Duffett-Smith 128.

[58] Comets P/ Macholz and P/Mellish, Marsden pp. 79, 82.

[59] As noted, for instance, by Pingré 277 and Rose 260 n. 90, and discussed above p. 63. See Appendix V for a tabular list of parallels between the *Memoirs* and our other sources, one of which (Servius, Text 6) makes direct reference to the *Memoirs.*

[60] See, Drumann-Groebe 773, "Stundentafel": these times are "Local Apparent Solar Time", where noon occurs at exactly 12:00. Our modern calculation gives, in "Local Mean Solar Time", the 11th hour as 4:59-6:14 P.M., with noon at 12:02 P.M.

[61] Baebius Macer is known only from two citations in Servius, Text 6 and on *Aen.* 5.556, which concerns a practice of the emperor Augustus at celebrations of the *lusus Troiae.* Since the *lusus Troiae* is likely to have been performed at the games of 44 B.C.

discrepancy.) Perhaps under the influence of Baebius' assertion, Servius twice mistakenly [?] reports that the comet was visible "at midday" (*die medio*, Texts 6, 7), and twice more loosely states that it was seen "during the daytime" (*per diem*, Texts 5, 8).

2. It was visible "over the course of seven days" (*per septem dies*, as reported by Augustus, Text 1; cf. Texts 9, 12) and then "vanished from sight" (ἠφανίσθη, Text 10). That is, it was seen from the 11th hour of the day "for seven nights" (ἐπὶ νύκτας ἑπτά, Text 10). (It is presumably this comet that lies behind Pliny's assertion [*NH* 2.90] that seven days was the shortest span during which a comet had been observed [see p. 69 for discussion].) A shorter duration of "three days" (*per triduum*) is given only in Servius Auctus (Text 5) and may concern the period of maximum brightness (see p. 130). The statement in one passage of Dio (Text 11) that a "new star" (which was most likely Caesar's comet) was visible "for many days" (ἐπὶ πολλὰς ἡμέρας) appears to be a red herring. For the correct interpretation of that passage, see p. 92.71. In another passage, Dio (Text 3) generalizes [?] in stating that the comet was visible "throughout all the days" of the games (παρὰ πάσας τὰς ἡμέρας ἐκείνας), which in imperial times occupied eleven days, 20 - 30 July. Although the celebration in 44 may have been on a more modest scale and of shorter duration than under the empire, it is nonetheless likely to have lasted for at least eight or nine days,[62] and the fact that Augustus (Text 1) does not equate the period during which the comet was visible with the length of his games points to the conclusion that it was not co-extensive with the festival as a whole.[63] Earlier (p. 72), we pointed out that the comet most likely experienced an anomalous brightening in late July, and if so, that could explain why it suddenly became visible during the daytime and then just as suddenly faded from sight at the end of only seven days. For the sake of argument, we shall assume that the comet was observed during 20–26 July, on three of which days (24, 25 and 26) we can be sure that the comet was seen, although we don't know whether it appeared during the first, second, third, fourth, or fifth seven-day

(see p. 55.47), the two fragments of Baebius (*HRR* II 71-2) were both conceivably drawn from his account of those games. Bardon (99) speculates that Baebius wrote a biography of the emperor Augustus and was his contemporary (so too Bramble 492); Bardon rejects the identification of Baebius Macer with Ovid's friend Macer (see G. Wissowa, *RE* 2.2 [1896], 2731, Baebius [31]).

[62] See above, pp. 47.23, and pp. 54-55.

[63] Hence we reject the conclusion of Halley 902 and Pingré I 277-78 that our sources give seven days as the period of visibility simply because that was the length of the festival.

period that can be fit into the maximum allowable span of eleven days (20–30 July).

3. According to Augustus' own words (Text 1, cf. Texts 3, 4, 5), the comet was observed *in regione caeli sub septentrionibus,* "in the region of the sky beneath the *septentriones* [viz., the seven bright stars in Ursa Major, which form the Big Dipper]". We interpret this expression to mean "in the northern region of the heavens" (not "under the constellation Ursa Major"), and we regard *sub septentrionibus* as equivalent to the adjective *septentrionalis* ("northern"), a rare word until the first century A.D.[64] Indeed, the expression *sub septentrionibus* is attested as meaning "in the north" elsewhere (e.g., Caes. *B.G.* 1.16.2; Gell. 9.4.6; Vitr. *Arch.* 6.1.3)—*septentriones* being the name given to one of the four *regiones caeli* (Gell. 2.22.3)—and significantly at least two of our ancient sources appear to have understood *sub septentrionibus* in Augustus' account as "in the north" (Text 3: ἐκ τῆς ἄρκτου, and Text 5: *in septentrione,* see nn. ad loc.).

Some scholars, however, have interpreted *sub septentrionibus* literally to mean that the comet was seen "beneath the Big Dipper".[65] Certainly the Latin will bear that meaning, but aside from the fact that our Greek and Latin sources (in contrast with the more detailed accounts from China) *very* rarely give the position of comets relative to the constellations (instead, usually telling us merely that a given comet first appeared in the North, South, East,

---

[64] The adj. *septentrionalis* is attested only in Varro (*Ling.* 9.24, *Rust.* 1.2.4, 1.24.3) before the first century A.D. but is quite common later in writers such as Seneca, Pliny and Columella (more than 100 citations in the texts contained on the Packard Humanities Institute CD ROM 5.3 "Latin Texts" databank). So, for instance, the 4th cent.(?) A.D. writer Obsequens (54) reports that in 91 B.C. a fire ball flashed "from the *northern* region" (*a septrionali regione*), while Orosius (5.18.3) in describing the same phenomenon states that it flashed "from the region *of the north*" (*a regione septentrionis*).

[65] E.g., Halley 901-3, Pingré I 278, Baldet 16, and Hasegawa 65. Boll 122-24 n.3, who assumed this location, tried to connect the comet with a constellation "Caesar's Throne" (*Caesaris Thronus*), which is attested only by Pliny (*NH* 2.178) and is, according to Boll, possibly to be identified with the representation on the Farnese Globe of what appears to be a chair with a high back. The trouble is, the passage in Pliny contains several major blunders, and since none of the Augustan poets or later Greek astronomers mentions "Caesar's Throne", it is difficult to believe that it had anything to do with such a celebrated event as the *sidus Iulium.* Gundel 1153.48-52 accepts Boll's conclusions, but see R. Böker, *RE* suppl. 8 (1956), 918-19, for a summary of different interpretations that have been offered before and after Boll. For illustrations of the Farnese Globe, showing what is possibly the outline of a throne, see Thiele 27 and Taf. III.

or West),[66] serious problems arise if we adopt the view that Caesar's comet was seen in the region of the sky below the Big Dipper.

First, since in 44 B.C. the Big Dipper extended from 8.3 to 12.3 hrs of R.A., and the R.A. of the Sun was very nearly 8.0 hrs. on 23 July (the midpoint of our seven-day period commencing on 20 July), if the comet had been located beneath the Big Dipper, it would have been in the NW at the 11th hour when it is said to have become visible (see Figure 3).

This would put the comet in a region of the sky that was growing brighter as the Sun drew closer to the western horizon, and it is difficult to explain why the comet was not seen much earlier in the day, as well as at the 11th hour.[67] Although daylight comets are sometimes observed quite close to the Sun (usually when they are near perihelion and have grown increasingly bright during their approach to the Sun[68]), they are usually visible throughout a good part of the day, even at high noon. By contrast, not only was the daylight comet of 44 not seen, according to our sources, until the 11th hour (approx. 5–6:15 P.M.), but at that time of day, if it had been situated beneath the Big Dipper, conditions should have been less, not more, favorable for observing it. On the other hand, if the comet appeared to the east of north (rather than to the west of north), its distance from the Sun could account for why it did not become visible until late in the afternoon, when the Sun was in the west. Furthermore, it could have first appeared at the 11th hour, and not earlier, because prior to that time it may have been below the horizon, if it rose in the NNE.

There is, in fact, some evidence pointing to this conclusion, and this evidence provides a second argument against locating Caesar's comet under the Big Dipper, in the NW. Not only did Augustus himself specifically

---

[66] Leaving aside the comet of 44 B.C., the positions of only three comets, out of the 62 listed in Barrett's catalogue, are given relative to other heavenly bodies by our Greco-Roman sources: no. 5 in 373/2 B.C., "rose to Orion's belt" (Aristot. *Mete.* 1.6.343b, text not quoted by Barrett); no. 44 in A.D. 60(?) near Boötes (*Octavia* 232) and the Corona Borialis (Lydus, *De ost.* 10b, text not cited by Barrett); and no. 57 in A.D. 389 near Venus (Philostorgius, *Hist. eccl.* 10.9).

[67] Pingré I 278, admitting that the comet would have been above the horizon for a considerable time before the 11th hour, if it was situated under the Big Dipper, and furthermore that it would have been well west of the celestial meridian, found these circumstances troubling, the more so because he assigned the comet to 23 Sept., when the R.A. of the Sun was 11.8 hrs. By that date in Sept., the comet would have been to the NW of the Sun if it was in the region of the sky under the Big Dipper.

[68] The three daylight comets observed in this century (comet 1910 I, comet 1927 IX, and 1965 VIII Ikeya-Seki) became so bright just prior to, or after, perihelion that they could be seen in broad daylight, attaining estimated magnitudes of -4, -6, and -10 respectively (Kronk 106, 122, 172).

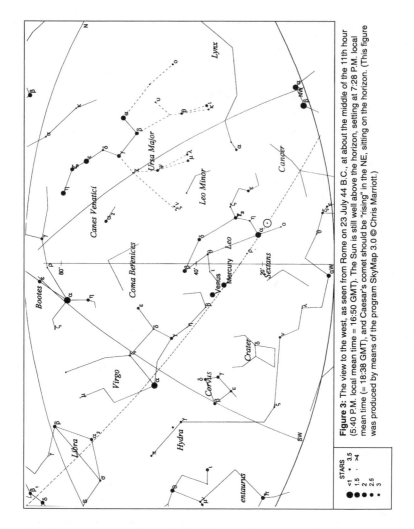

Figure 3: The view to the west, as seen from Rome on 23 July 44 B.C., at about the middle of the 11th hour (5:40 P.M. local mean time = 16:50 GMT). The Sun is still well above the horizon, setting at 7:28 P.M. local mean time (= 18:38 GMT), and Caesar's comet should be "rising" in the NE, sitting on the horizon. (This figure was produced by means of the program SkyMap 3.0 © Chris Marriott.)

describe the comet as "rising" (*oriebatur*, Text 1; cf. Texts 4 *exorta*, 9 *exoriens*, with the variation *emersit* in Text 2), but so too, apparently, did one other contemporary observer, Baebius Macer (Text 6 *ortam*). This agreement of two eyewitnesses is all the more striking because the verb *orior* ("rise") is very seldom applied to comets, which are more often said to "appear" (*appareo*), "be seen" (*videor*), "gleam" (*effulgeo*) or "blaze" (*ardeo*).[69]

---

[69] The verb *oriri* is found in the descriptions of only two other comets in Barrett's catalogue of 62 comets, both of which are likely to have appeared in the E: (1) no. 45, the Neronian comet of A.D. 64(?): since it was seen "throughout several nights in a row" (*stella crinita . . . per continuas noctes oriri coeperat*, Suet. *Nero* 36.1), it most probably

Furthermore, since we know that the comet of 44 was seen late in the day, in the 11th hour, and Plutarch (Text 10) states that it was visible "for seven nights" (ἐπὶ νύκτας ἑπτά), it presumably had to either be circumpolar or rise in the NNE in order to remain above the horizon throughout the night. A position beneath the Big Dipper might fulfill the circumpolar requirement, but, as we have argued, that location does not suit an object that remained invisible until the 11th hour, and it could scarcely be said to "rise" at that time of day, in that region of the sky.

Faced with this difficulty, Pingré (I 278) felt obliged to reject the literal meaning of *oriebatur* ("rose"), taking it instead as "became visible" (*apparebat*), while Halley (901-3) sought to solve the dilemma by having the comet "rise" during the "11th hour of the *night*" (emending *diei* in Text 1 to *noctis*, or omitting it altogether as in Text 9). In doing so, Halley failed to realize that the "11th hour of the *day*" (*undecimam horam diei*) must have been what Augustus wrote because (1) the identical expression is adopted by Seneca to give the rising time of Comet Caesar (Text 2) and (2) several other sources (Texts 5-8) make it plain that Comet Caesar was a daytime, not a predawn phenomenon (the 11th hour of the night = approx. 3:55–4:40 A.M.). Yet even *with* his emendation, Halley still could not succeed in making the comet "rise" under Ursa Major "at the 11th hour of the night" at the time of Octavian's games, which Halley assigned to 23 September Halley, therefore, had to fall back on the argument that when the comet was seen under UMa, it had experienced a retrograde motion from the sign Virgo to Cancer, but that assumption, of course, puts the comet precisely in the vicinity of the Sun, which was just to the west of the star Regulus (α Leo) on 23 July (see Figure 3).

Our interpretation of *sub septentrionibus*, by contrast, has the advantage of making the best sense of the available evidence, and we conclude, therefore, that the comet was seen somewhere between NNE and due N in that region of the sky bounded by declinations 43.4° to 48° at the latitude of Rome, and by the 11th and 12th hour rising lines. This would put the comet just south of Cassiopeia, and north of Perseus and Andromeda, a region which, incidentally, lies just south of Ursa Minor, a constellation that is

---

rose in the E and appeared to move across the sky, setting in the W, although we cannot rule out the possibility that *per* + the pl. *noctes* may refer to the period of time represented by several successive nights rather than to each individual night; (2) no. 21 in Barrett (Justin 37.2.3, *cum oreretur occumberetque, IV horarum spatium consumebat*, text not quoted fully by Barrett), which we hope to show elsewhere should be dated to 135, rather than 134 (so Barrett), and most likely corresponds to a "long-star" (*ch'ang hsing*) that according to our Chinese sources "appeared in the east, stretching across the heavens for 30 days" (Ho no. 39).

sometimes called *septentrio minor* (*OLD* s.v. 1b), though it is more often
known by the name Cynosura, the "Dog's Tail" (*OLD* s.v. 1). Could it be
that a reference to Ursa Minor by its less familiar name (traces of which may
lurk in Text 4: *sub septentrionis sidere*) became displaced by the substitution
of *septentrionibus* for *septentrione minore* because *minore* perhaps dropped
out of the text at some stage, or was overlooked by some reader or copyist?

Be that as it may, both allowed regions are shaded in Figure 4, the
Chinese to the left, and the Roman to the right. The regions are connected by
the path of a comet having a perihelion distance of 0.224 A.U. (See Table 1, p.
126, for the orbital parameters).

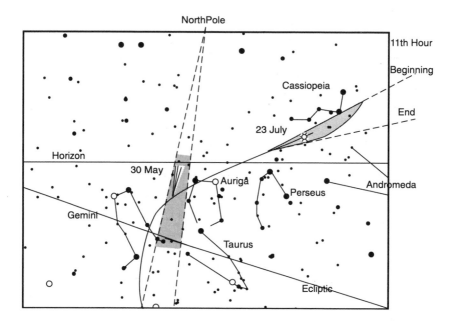

Figure 4: The view to the north, as seen from 42° N (the latitude of Rome rounded to the
nearest whole degree), at sunset on 23 July 44 B.C. The allowed regions for the comet's
positions are shown shaded. The region on the right is bounded by a curve showing those
points that rise at the NNE, and by two straight lines, one showing those points that rise at
the beginning of the 11th hour, and another showing those points that rise at the end of the
11th hour. A path is plotted for a retrograde orbit with perihelion = 0.224 A.U. This path
shows the comet within the allowed region on 23 July.

Intermediate Sightings?

When the Chinese observed Caesar's comet in May-June (approximately days -6 to -2 in Figure 2, p. 82), it was probably very close to perihelion, and therefore at its brightest before its much later outburst in late July when it briefly became daylight visible. It is possible that the Chinese then lost sight of the comet for several days, when it moved into the skyglow (days -1 to +1 in Figure 2). As seen in Figure 2, after perihelion the comet is likely to have moved deeper into the western skyglow, until it finally set at the same time as the Sun on day +8. Therefore, it is unlikely to have been visible in the evening in the NW for more than a day or two after day 0. During this same period, it should have emerged from the eastern skyglow so as to be visible at an ever increasing interval before sunrise (just over an hour before sunrise on day +2; approximately two and one half hours before sunrise on day +8). It would, however, have been farther away from the Sun when seen during the predawn, and its distance from the Sun would have been increasing with each passing day.[70] Therefore, during the predawn hours the comet may have been dimmer than it was when the Chinese saw it in the evening twilight in May-June (roughly days -6 to -2), and its tail may have grown smaller. Its apparent magnitude depends on both its distance from the Sun and its distance from the Earth. We show in chapter six that it most probably faded gradually after its initial sighting in May-June. It then apparently experienced in July a brief and very intense flare-up, which caused it to be visible in broad daylight.

Possibly during this intermediate period, between the Chinese sighting in May-June and the Roman report in late July, the comet was, in fact, too dim to be seen with the naked eye. This would account for the apparent failure of our sources to mention the comet during the intermediate period. As we shall see below (pp. 119-24), however, our estimates of the comet's brightness before the flare-up in July point to a decrease in magnitude from perhaps -3 or -4 at perihelion in late May to a range of +1 to +5 during the latter part of June and the first three weeks of July. Therefore, it should have been visible, and there is, in fact, a hint in two of our texts (nos. 11 and 12) that an observer in the West may have caught sight of Caesar's comet in the predawn hours, or during the night, in June or early July. These texts recount that a "torch" (λαμπάς, *fax*) was observed to move from east to west and that a "new star" was seen for a period of days. The "new star" is almost

---

[70] See Figure 8 (p. 129) for estimated distances of the comet from the Sun on 30 May and 23 July.

certainly the *sidus Iulium*, as we can tell from the further statement in one of these two sources (Text 12) that it was seen "for seven days" and was "clearly visible" (*insignis*).[71]

Scholars have tended to overlook or misinterpret the evidence found in these two passages because they have been misled by the context of Dio's account (Text 11). It forms part of a digression on portents that foreshadowed the outbreak of war in 43,[72] and although it is inserted after Dio's account of several meetings of the Senate in January 43, the list is clearly intended to summarize the portents that occurred in 44, as we can tell from the nearly identical list of portents in Obsequens under the year 44.[73] The striking similarity between these two lists makes it clear that Dio and Obsequens must have been drawing upon a common source (Livy or some intermediary?), one that provided a summary of portents in 44 without undue emphasis on any one of them. The account given by their common source is likely to have resembled the lists of portents in 44 that are found in Virgil and Tibullus (from which Texts 31 and 32 are drawn). These lists treat the comet as a baleful omen, having no connection with the *sidus Iulium* that appeared during Octavian's games, just as if there were two distinct phenomena rather than a single comet which some (Octavian and his supporters) interpreted as a sign of Caesar's apotheosis, while others (Octavian's opponents) interpreted it as a sign of impending war. Therefore, although both Dio and Obsequens had treated the *sidus Iulium* separately, earlier in their works

---

[71] Pingré I 277 failed to appreciate the significance of this crucial bit of evidence in Obsequens (Text 12) and mistakenly took Dio's less precise statement that "a new star was seen for *many days*" (Text 11) as proof that Caesar's comet was visible for longer than the period of 7 days to which the bulk of our other sources refer (see p. 85).

[72] Weinstock 370 n. 3, for instance, correctly notes that Obsequens twice refers to Caesar's comet (Texts 4 and 12); but he misinterprets Dio's report (Text 11), which is virtually identical to Obsequens' (Text 12), as mentioning "the comet again among the evil signs of 43 B.C." (p. 371 n. 5). Weinstock apparently failed to appreciate that Dio was recounting portents that occurred in 44, not in 43, in a digression from his narrative framework. In the two most recent catalogues of comets in Greco-Roman sources, Gundel 1187 neglects to take Text 12 (Obsequens) into account and assigns Text 11 (Dio) to 43 B.C. (interpreting it as foreshadowing Cicero's death), while Barrett 96 reproduces Gundel's error in assigning Text 11 to 43, yet lists Text 12 where it belongs under 44 (surprisingly omitting the earlier and fuller passage in Obsequens, Text 4!).

[73] For instance, both lists contain the following portents in this order: (1) frequent strikes of lightning (cf. Hor. *Carm*. 1.2.2-4); (2) a statue dedicated by Cicero on the Capitoline before his exile was overturned and shattered by a windstorm; (3) three suns were seen, one of which was surrounded by a fiery crown resembling spikes of grain; (4) letters in the names of Antony and Dolabella were damaged on a tablet in the temple of Castor; (5) dogs howled outside the house of Lepidus, the *pontifex maximus*; and (6) the Po River overflowed its banks and left behind numerous snakes when the water receded.

with fuller details (Texts 3 and 4), they were probably unconscious of the repetition when they referred to the same comet again among the ominous signs of impending war.

What is of the greatest interest for our present purposes, however, is the reference in both texts to a "torch", in contrast with the "new star." Several interpretations are possible, and unfortunately there is no way of deciding which is most likely to be correct. The "torch" could simply be a meteor or shooting star. In that case, the texts have no bearing on the sighting of the *sidus Iulium*.[74] On the other hand, *fax* and λαμπάς are both sometimes used to denote a comet generally, or a comet having a particular shape.[75] Obsequens only once refers to a comet as a *stella crinita* (in his earlier description of the comet of 44), and elsewhere employs *fax* in at least eight other passages. Roughly half of these reports are likely to have concerned a comet;[76] about the rest we cannot be certain because we have no corroborative evidence for a comet in other sources. The same holds true in evaluating Dio's use of λαμπάς. Although, in contrast with Obsequens, Dio quite often employs the term κομήτης, it is clear that on a number of occasions λαμπάς is likely to denote in Dio a comet rather than a meteor.[77]

If we interpret the "torch" in Texts 11 and 12 as a comet, we are still faced with at least two mutually exclusive interpretations of the reports. On the one hand, the two texts may simply allude to the well-attested dispute that arose in late July over whether the bright light in the sky was a "new star" or a comet (Texts 3 and 6); on the other hand, the texts may concern a sighting of Caesar's comet during the interval that fell between the Chinese observations in late May and the Roman observations during 20-30 July. In

---

[74] For instance, Cicero specifically contrasts *faces caelestes* with what the Greeks called *cometae* (*Nat. D.* 2.14), and the *faces* in Livy 41.21.13 for the year 174 B.C. must be meteors: "on that same night, a considerable number of *torches* glided downward through the heavens" (*faces eadem nocte plures per caelum lapsae sunt*). λαμπάς most likely describes meteors in [Aristotle], *Mund.* 395b11, and Dio 47.40.2, 60.19.4.

[75] *Fax* is Manilius' term for a comet (1.867, 893); it is applied to a comet in [Sen.] *Octav.* 232; and it possibly refers to a comet in Lucretius 2.206 and Cic. *Cat.*, 3.18. *Lampas* is clearly a comet in Calp. Sic. 1.81 (Text 33), and λαμπάς almost certainly describes a comet in Diod. Sic. 15.50.2, 16.16.3. A particular kind of torch-like comet was called *lampas* (Manilius 1.846, Sen. *Nat. qu.* 1.15.4) or *lampadias* (Pliny, *NH* 2.90; Serv. on *Aen.* 10.272).

[76] In §§24, 51, and 71 under the years 137, 94, and 17 B.C. A comet is attested for each of those years in sources other than Obsequens (see Barrett nos. 20, 25, 39), though the evidence for 17 B.C. is subject to dispute (see p. 65.12).

[77] Almost certainly in Dio 50.8.2 (a torch seen for a period of many days in 37 B.C.) and possibly 54.19.7 (a torch moved throughout the night from S to N in 17 B.C.): see Barrett nos. 37, 39.

the latter case, the movement of the "torch" from east to west will have described its motion through the constellations (see Figures 4 and 8, pp. 90 and 129). This movement would have been about one degree each day, and could have been quite noticeable over the span of a week. Some slight support for the notion that Texts 11 and 12 may concern intermediate sightings of Caesar's comet is found in Virgil (Text 31), who writes that "at no other time [as after the assassination of Caesar] did fearsome comets so often blaze". Unless we dismiss this reference to the appearance of numerous comets as poetic exaggeration,[78] or the compression of several years' events into one,[79] we may see behind Virgil's words a possible allusion to sightings of the same comet during three different periods: (1) in May-June (in the NW just after sunset), (2) in June-July (rising in the NE at an ever increasing interval before sunrise, and moving gradually westward through the constellations), and finally (3) in late July (rising in the NE before sunset, at the 11th hour). These sightings could easily have been mistaken in antiquity as manifestations of more than one comet, just as Newton at first treated the morning and later evening appearance of the comet of 1680 as two different comets.[80]

---

[78] Clearly Ovid refers to the same phenomena, whether real or imagined, in *Met.* 15.787: "often torches were seen to blaze among the stars" (*saepe faces visae mediis ardere sub astris*). In Ovid, the description figures in a list of portents that foreshadowed the Ides of March; the passage in Ovid recalls Virgil's account in the *Georgics* of the portents that supposedly occurred *after* the Ides.

[79] The statement may reflect the tradition that several comets were seen during the years that intervened between the Ides of March in 44 and the Battle of Philippi in 42 (Manilius 1.907-908).

[80] See above, 69.27.

V

THE TROUBLING SILENCE OF OUR SOURCES

We now have a rough position for the comet for several days sometime during the latter part of May and the first part of June, and another around 20 to 26 July. Before we attempt to reconstruct a probable orbit for Comet Caesar, let us first consider what appear to be some surprising gaps in our sources. On the one hand, our Greco-Roman sources fail to confirm the Chinese sighting of a comet in the spring of 44 B.C., and on the other hand, both the Chinese sources and the contemporary letters of Cicero fail to attest the sighting of a comet during late July when Octavian held his games. This silence of our sources is bound to arouse suspicion and cast a shadow of doubt over the reliability of the sources that *do* report the comet of 44 unless some logical explanation can be offered for the absence of reports in both the East Asian and European sources. Let us take a closer look at this potentially negative evidence and try to discover how much weight should be given to it.

THE SIGHTING IN MAY

Our Western sources offer no account of the comet in the spring of 44 comparable to the report from China, despite the fact that there are extant a good number of letters written by Cicero and his friends during the very period of the Chinese sighting (18 May-16 June). In all, there survive 21 letters for this span of 30 days, eighteen written by Cicero to Atticus (nos. 378-95 in Shackleton Bailey) and three in the collection *Ad Familiares* (nos. 328-30 in Shackleton Bailey), but not one of those letters contains the faintest allusion to the comet reported by the Chinese in the NW, soon after sunset. At first glance, this silence may appear ominous, and in other circumstances it might well cause us to question the historical reality of a comet that is made to follow so closely upon the heels of the assassination of

Caesar. In this instance, however, there is no reason to suspect that the Chinese sources invented the comet. Certainly they had no awareness of Caesar's death in the West, and as we have demonstrated, those sources tend to be very reliable (pp. 66-67). Therefore, the problem raised by the silence of our Western sources calls for a solution that will account for their failure to take note of a celestial phenomenon whose historical reality is scarcely to be doubted.

If we begin by taking into consideration Cicero's personal view of such phenomena as comets, then the absence of any mention of a comet in the letters written in the spring of 44 need not be all that puzzling. Even supposing that Cicero had caught sight of the comet when it was observed from the Chinese capital, it is unlikely that he would have read any great significance into the celestial object because he was not of a superstitious nature with regard to such manifestations. In his treatise, *De Divinatione*, to which he added the finishing touches shortly after the Ides of March in 44 B.C.,[1] Cicero had resoundingly rejected the notion that showers of stones and blood, shooting stars, comets and the like were to be viewed as supernatural signs of things to come (*Div.* 2.60).[2] Therefore, if he had chanced to observe our comet in the spring of 44, it would have held no particular significance for him personally, and so we would hardly expect him to treat it as worthy of remark in his letters.[3]

By contrast, the silence in another body of evidence from the Greco-Roman world is less easy to explain. We possess in both prose[4] and poetry[5] quite lengthy lists of portents that supposedly occurred in 44, but not one of those lists contains a clear and unambiguous reference to a sighting of the comet in the spring of 44 (May–June), as opposed to the later sighting of that same comet in July, during Octavian's games.[6] One would expect the

---

[1] Schanz-Hosius I 515.

[2] See now J. Linderski 36-38 for a convincing explanation of why in later life Cicero came to reject divination as an institution worth defending, although less than a decade earlier he had adopted a contrary position in his treatise *De Legibus*.

[3] In an earlier draft of this chapter, we were inclined to read greater significance into the silence of Cicero's letters from May and June than is warranted. We thank Alexander Jones for discussing this question with us and stimulating us to re-examine some of our basic assumptions.

[4] Dio 45.17; Obsequens 68; add possibly Appian 4.4.15, whose list of portents marking the formation of the so-called Second Triumvirate bears a striking resemblance to the lists given by Obsequens and Dio for 44 (see below p. 141 for discussion of this point).

[5] Verg. *Georg.* 1.463-88; Tib. 2.5.71-78; Ov. *Met.* 15.782-98, clearly based on Virgil, assigns the portents to the period immediately leading up to the Ides of March.

[6] See p. 94 for a discussion of the possibility that Virgil alludes to the spring sighting in Text 31.

comet to have been mentioned in those texts, if it had been widely observed by the Romans in the spring of 44, since according to a prevailing popular notion a comet was often to be interpreted as a harbinger of war (see p. 135), and in that period the mood of the public was increasingly dominated by the dread of renewed warfare. We can tell from Cicero's correspondence during the spring of the year that he and his friends came to view the renewal of civil war as inevitable.[7] The extant lists of portents that are said to have led up to, or followed, Caesar's assassination capture something of the spirit that must have existed among the common people at the time. Therefore, if a comet had been sighted and had been plainly visible so soon after the Ides of March, in May or June of 44, it would surely have caused a stir at the time, at least among the common folk. Writing nearly a century later, Seneca goes so far as to assert that the appearance of a comet inevitably arouses curiosity and heightens forebodings in persons who normally pay little heed to the stars when the heavenly bodies follow their normal courses.[8] Given these circumstances, it is certainly odd that our Greco-Roman sources contain no notice of the comet in May-June, comparable to the report from China. This silence of our western sources cries out for an explanation, and a number of factors are worth considering.

Part of the explanation may lie in the fact that the Chinese conducted their observations from a tower, using instruments, and drawing upon years of training and experience that might permit them to take note of objects that could go undetected by a casual observer.[9] Indeed, the absence of any

---

[7] The recruiting activities of Mark Antony among Caesar's veterans in Campania in late April, early May, caused Cicero to write on 11 May, "I'm convinced the matter points in the direction of armed conflict," (*mihi autem non est dubium quin res spectet ad castra*, *Att.* 14.21.3). Cicero's apprehension was shared by Balbus (*Att.* 14.21.2), Hirtius (*Att.* 15.1.3), and Brutus and Cassius (*Fam.* 11.2). Another source for concern was the prospect that Sextus Pompey would precipitate a renewal of civil war by leading his army from Spain into Italy (*Att.* 14.22.2 of 14 May). Later (during the period of the Chinese sighting, 18 May - 16 June), Cicero remarked on 24 May, "it seems to me that his [Antony's] whole policy looks to war," (*mihi totum eius consilium ad bellum spectare videtur*, *Att.* 15.4.1); and on 15 June Cicero wrote, "it seems to me that conditions point to bloodshed that will soon be upon us," (*mihi res ad caedem et eam quidem propinquam spectare videtur*, *Att.* 15.18.2).

[8] "The same thing happens in the case of comets. If there appears a flame that is seldom seen, one of unaccustomed shape, everyone is eager to know what it is. Without regard for the other heavenly bodies, everyone seeks to learn about the newcomer and is uncertain whether he ought to feel awe or dread." (*Idem in cometis fit: si rarus et insolitae figurae ignis apparuit, nemo non scire quid sit cupit et, oblitus aliorum, de adventicio quaerit, ignarus utrum debeat mirari an timere, Nat. qu.* 7.1.5).

[9] See p. 66.16. Dubs III 553 remarks that many of the eclipses recorded in the *HS* would have had a magnitude too small to have been observed by the average person,

account of the comet in our western sources for the period of the Chinese sighting (May-June) strongly suggests that the object was not plainly visible at that time of year, and the reason why this may have been so is not difficult to imagine. As we have remarked earlier, when the comet was first observed by the Chinese in the evening sky, in the NW, it most likely remained rather deep in the sunset skyglow (p. 83). This circumstance could readily account for the failure of the comet to attract much notice on the part of the Romans. It may also be relevant that in contrast with the interest shown by the Chinese and Babylonians in observing the heavens and recording their observations, the Roman state never established a comparable astronomical bureau, nor were even educated Romans much inclined to make direct observations of physical phenomena for their own sake. For instance, even Seneca, who wrote his *Naturales Quaestiones* late in his life (circa A.D. 62-64)[10] and devoted the whole of book seven of that work to comets, takes into account only two comets in his own age (7.23.1 *duo nostra aetate*), although it appears likely that as many as five were visible to the naked eye from A.D. 54 to 64, the year before Seneca's enforced suicide in April 65.[11]

On the other side of the coin, however, it must be admitted that in antiquity the average person was far more accustomed than his modern counterpart to pay close attention to the night-time sky. For one thing, in the absence of modern time-keeping devices, the stars will have played an important role as a means of telling the time of night and of keeping track of the advancing seasons of the year.[12] For another thing, there was no industrial pollution and there was no glare of artificial illumination given off by

---

thereby demonstrating the care with which observations and calculations were made at the imperial observatory. Schove's "Comet Check-List" covering the 1st cent.–10th cent. A.D. (Appendix B, 285-97) demonstrates that the same holds true for comets. Of the 31 sightings that are classified by Schove as "Not noted by the general public" or "Noted only by experienced sky-watchers", all but one are attested in the Chinese sources, and only 5 of these 30 sightings caught the eye of observers outside of China (2 seen also from Japan; possibly 3 seen from Europe).

[10] Schanz-Hosius II 700; Oltramare I vi-vii.

[11] Seneca's two comets are those of A.D. 54 and 60 (Barrett nos. 43 and 44). He fails to take any notice of the comets of 55/6, 61 and 64 (Ho nos. 71, 73, and 75), but the absence of the latter may be explained by assuming that it appeared after the completion of Seneca's treatise. See Rogers 241, who concludes that Seneca was not a very keen observer but rather a "bookish scholar."

[12] It seems, for instance, that Caesar's poem *De astris* ("Concerning the stars") was intended to provide the Italian husbandman with a reliable guide for regulating his activities according to the risings and settings of the various constellations as observed from the latitude of Italy, whereas previously such astronomical-meteorological calendars (παραπήγματα) reflected calculations made for more southerly regions by the Chaldaeans, Egyptians, and Greeks (Pliny, *NH* 18.211). See *RE* 3A (1927) 1154.

large urban centers to interfere with a clear view of the stars. All of these factors should have insured that people would have been more prone in the first century B.C., than they are today, to notice anything unusual in the night sky. How then are we to account for the surprising fact that our western sources neglect the comet of 44 until late July, when it suddenly emerged in broad daylight?

Eruption of Mt. Etna

In our opinion, the most promising answer to this question is furnished by evidence pointing to the presence of a volcanic dust veil in the atmosphere over Italy in the spring of 44, precisely at the time when the comet was seen from China but did not attract much, if any, notice in Italy. [13] We may even be able to identify the probable origin of this volcanic aerosol, although certainty is impossible because we lack a world-wide record of volcanic activity for this early period. The leading candidate as the source of the aerosol is Mt. Etna, which, according to a notice attributed to Livy, gave off such a great blast of fire before the murder of Caesar that the heat was felt across the Strait of Messina in southern Italy. [14] Although it may seem an incredible coincidence that both an eruption of a volcano in the Mediterranean and the apparition of a comet should occur in the same year, on either side of the date of Caesar's

---

[13] The supernova of A.D. 1054, whose remnant today forms the Crab Nebula in Taurus, may provide a parallel instance: there are no reports of it from Europe or Arab North Africa, but it was seen over the course of more than a year from China and was daylight-visible for some weeks. It is also attested by rock paintings and carvings from the American Southwest and possibly by a coin from the Middle East. Ridgway (a work known to us by précis) speculates that the eruption of an Icelandic volcano may have produced unfavorable viewing conditions from Europe.

[14] Livy 116 fr. 47 (ap. Servius on *Georg.* 1.472): "It is a bad portent when Mt Etna of Sicily emits not puffs of smoke but balls of flame; and as Livy reports, such a quantity of flame poured forth from Mt Etna before Caesar's death that not only the neighboring cities but even the community of Regium, which is some considerable distance away, felt the blast of heat." (*Malum enim omen est, quando Aetna, mons Siciliae, non fumi, sed flammarum egerit globos; et ut dicit Livius, tanta flamma ante mortem Caesaris ex Aetna monte defluxit, ut non tantum vicinae urbes, sed etiam Regina civitas, quae multo spatio ab ea distat, adflaretur.*) Regium, on the toe of Italy, across from Sicily, on the Strait of Messina, lies well beyond the region normally affected by Etna. Usually ash fell to the north not much beyond the Sicilian coastal town of Tauromenium, ca. 18 Roman miles to the NE (Pliny, *NH* 3.8), but on occasion Etna is said to have hurled flaming balls of hot sand up to 50 or 100 Roman miles away (Pliny, *NH* 2.234)—Regium being well within this range, just under 50 Roman miles to the NE of Etna. Simkin-Siebert 40 assign to the eruption in 44 an estimated magnitude of 3 on the VEI scale (Volcanic Explosivity Index, ranging from 0 to 8 in increasing explosivity), 3 = explosive and severe.

death,[15] the historical reality of the comet, as we have seen, is established beyond all reasonable doubt. So too, an explosive eruption of a volcano in the spring of 44 (quite likely Mt. Etna) can now be shown to be virtually an assured historical fact thanks to a stream of corroborative scientific data which have been developed in the last decade or two. In all, there are four separate strands of evidence that support the notices in our literary sources attesting an eruption of Etna in the spring of 44.[16]

First, there is present in our Greco-Roman sources a recurrent theme that the period following Caesar's assassination was marked by unusual solar and meteorological phenomena. For instance, the light of the Sun is said to have been feeble and pallid throughout the year.[17] Virgil (*Georg.* 1.466-68) describes the Sun as "covering his shining face with dingy gloom" (*caput obscura nitidum ferrugine texit*) after the death of Caesar to warn of the coming civil wars; as a result, the wicked age feared an "everlasting night" (*aeternam noctem*). A century later, Pliny (*NH* 2.98) appears to refer to the same weakened condition of the Sun in less poetic language when he states that after the murder of Caesar and in the following year there occurred portentous and lengthy *defectus solis*. Although the expression *defectus solis* on its own could mean "solar eclipses" (*TLL* V.1.292.33-59), it must be interpreted differently in this context because the phenomena are said to have produced a "gloom that lasted for the good part of a year" (*totius paene anni pallore continuo*) that stretched into 43 B.C. during the war with Antony (*Antoniano bello*). Moreover, in 44 B.C. no solar eclipse was visible from

---

[15] Sturt Manning in an e-mail communication writes: "Actually, comets may be part of a forcing mechanism that leads to eruptions. There is almost a significant correlation; but do not expect too many scientists to believe it." In our view, this apparent correlation is best explained by the fact that although volcanoes erupt with relative frequency (e.g., 46 known or suspected eruptions in the 1st cent. B.C., including Etna's in 44 B.C.: Simkin-Siebert 184) and naked-eye comets are not uncommon (roughly an average of 43 per century for the years A.D. 800-1700 [Yeomans (2) 115]), the ones that we hear about, because they make their way into the historical record, tend to be the those that happen to fall close to some major historical event (such as the assassination of Caesar). This process of selection inevitably has the tendency of bringing comets and volcanoes together in historical accounts, without there being, however, any causal link between them (see pp. 111-12 for discussion).

[16] In addition to Livy (quoted, p. 99.14), Virgil (*Georg.* 1.471-73) describes repeated, fiery outbursts of Etna *after* Caesar's murder, and possibly we should connect with this volcanic activity the shower of stones listed among the portents of 44 (Tibullus 2.5.72) and 43 (Appian 4.4.14). Surprisingly, however, the eruption of Etna is not included among the portents of 44 by either Dio (45.17.3-7) or Obsequens (68), who appear, from the striking similarity of their accounts, to go back to a common source (see p. 92).

[17] Ov. *Met.* 15.785-86; Tib. 2.5.75-76; Plut. *Caes.* 69.4; Dio 45.17.5.

Rome.[18] As scholars have rightly noted, the gloomy condition of the atmosphere that is so amply attested by our Greco-Roman sources for the year 44 bears all the hallmarks of the aftermath of a violent volcanic eruption, especially one so close to Italy as Mt. Etna.[19] Besides the pallor of the Sun's rays in 44, triple suns are said to have been observed, one of which appeared to be encircled by a corona, a phenomenon that has quite often been reported after major volcanic eruptions.[20] The tradition that the Sun exhibited this unusual appearance on the day on which Octavian entered Rome some months after the murder of Caesar permits us to assign these solar phenomena to the early part of May.[21]

It is difficult to estimate the extent of the fallout from the eruption. One source (Plut. *Caes.* 69.4) goes so far as to assert that the crops in 44 failed from lack of sunlight and warmth, and another source (Obseq. 69) reports a dire oracle attributed to Apollo at Delphi which foretold no harvest of grain in 43. These conditions are reminiscent of climatic changes that have taken place in the wake of exceptionally massive volcanic eruptions,[22] but we have to admit that in the Mediterranean, at least, the evidence for significant reductions in crop production is extremely tenuous.[23] Apparently Sicily itself

---

[18] See Appendix VII for a discussion of the evidence.

[19] The classic modern study of these sources and the probable link between the unusual solar and meteorological phenomena in 44 and an eruption of Etna is by Stothers-Rampino 6358-60, supplemented by Rampino et al. 88-9. Without reference to that scholarship, Mynors (on *Georg.* 1.469) drew the same conclusion, while Weinstock 282-83 less plausibly interpreted the references in our sources to a darkened Sun as no more than symbolism conveying the demise of Caesar's radiance.

[20] Obseq. 68; Dio 45.17.5; Jerome, *Chronicle* Olymp. 184.1. So, following the eruption of Krakatoa (August 1883), a volcanic dust veil moving at high altitudes caused a corona to be visible intermittently around the Sun from 5 Sept. 1883 to June 1886. This "Bishop's Ring" (taking its name from Sereno Bishop, who observed it from Honolulu on 5 Sept. 1883) varied in prominence, depending upon atmospheric conditions: Symons 232-63, summary 255-56. Likewise, the eruption of El Chichón in SE Mexico in 1982 produced a heavy dust cloud that lasted approximately two weeks and caused a prominent aureole to surround the Sun: Meinel 58. To judge from the reports in 42 of triple suns (Obseq. 70) or of a sun triple its normal size (Dio 47.40.2), the effect that the eruption of Etna in 44 had on solar phenomena may have lasted for quite some time, demonstrating that it was indeed an explosive eruption.

[21] Vell. 2.59.6; Sen. *Nat. qu.* 1.2.1; Pliny *NH* 2.98; Suet. *Aug.* 95; Dio 45.4.4; Obsequens 68, Lydus, *De ost.* 10b; Zonar. 10.13; cf. Livy, *Per.* 117. See p. 1.1 for the probable date of Octavian's entry into Rome in the spring of 44.

[22] For instance, the eruption of the Indonesian volcano Mt. Tambora in 1815 caused famine in Europe and such severe climatic changes as far away as North America that 1816 came to be known as the "year without a summer": Stommel (1983). We thank our colleague James Dee for directing our attention to Stommel.

[23] See Denniston on *Phil.* 1.13 and Garnsey 202 on *Att.* 14.3.1. Forsythe 53-55 tries

could be viewed as a potentially bountiful source of grain in June of 44,[24] and in the spring of 43 the production of grain in the vicinity of Cordova, Spain, whose latitude (37° 53'N) is near that of Etna's (37° 44'N), appears to have been quite normal.[25]

On the other hand, wholly independent reports from China do attest famine in 43 and 42, and they also attest various unusual solar and meteorological phenomena which are reminiscent of those recounted by our Greco-Roman sources.[26] For instance, the light of the Sun is said to have been weak and to have had a bluish cast in May and June of 43; the summer was cold, and the Sun did not recover its brilliance until later in the year (*HS* 27Cb: 17a). The harvest in 43 was apparently delayed by the cold and ultimately destroyed by an early frost in the autumn before it could be gathered from the fields (*HS* 27Bb: 15a). This resulted in famine, which continued into the following year *HS* (9:8a-9a). These conditions in 43-42 B.C. were presumably noted in the vicinity of the Chinese capital Ch'ang-an (34° 17'N) and so point to the probable existence of a volcanic veil at a latitude close to that of Etna's (37° 44'N). These accounts from China provide a welcome second strand of evidence corroborating the tradition in our western sources that Etna experienced an explosive eruption in the year of Caesar's murder.

Nature's own records of past volcanic eruptions yield the third strand of evidence that points to the presence of a volcanic dust veil in the year of Caesar's comet. These records are preserved in glaciers, and we are gaining more and more insight into the chronology of major volcanic eruptions by studying the deposits left in the polar ice caps by the sulfuric acid (H_2SO_4) aerosols that are produced by volcanic eruptions rich in sulfuric dioxide

---

to show that the winters of 44/3 and 43/2 were unseasonably cold and discusses evidence pointing to the possible existence of famine in 44 and a more serious food crisis in Egypt in 43-42, all of which, she argues, may have been caused by an eruption of Etna.

[24] Although the grain commission voted to Brutus and Cassius in June of 44 (cf. above, p. 45.10) may reflect some concern over impending food shortages, the cause of the anticipated shortages may have been political unrest following the assassination of Caesar, rather than adverse agricultural conditions. It should be noted that Cassius was to have been posted to Sicily (*Att.* 15.9.1), which implies that the island as a whole was not too adversely affected by an eruption of Etna. Apparently, too, the rich agricultural land in the Leontine Plain (approx. 48 km. to the south of Etna) escaped serious damage; Cicero repeatedly criticizes Antony for making a gift of that domain to his friends and cronies (*Phil.* 2.43, 84, 101; 3.22; 8.26), but there is no hint that the value of that land had in any way been lessened by a recent eruption of Etna.

[25] In the first half of June, Pollio wrote to Cic. (*Fam.* 10.33.5) *frumenta aut in agris aut in villis sunt* ("the grain is in the fields or on the farms", i.e., it has been cut and either lies in the fields or is already in the barns).

[26] For a good discussion of this evidence, see Bicknell 4-8.

(SO_2). Evidence found in ice cores taken from Greenland points to significant volcanic activity precisely in the decade when our literary sources tell us that Etna experienced an explosive eruption. Hammer et al. (233) found in the Camp Century ice core an acidity peak spreading over three years at a level corresponding to 50 ± 30 B.C., and in another ice core, Dye 3, Herron (3055-56) found a strong signal of volcanic activity occupying a five-year period which he assigned to circa 40 B.C.[27] Although scholars were quick to assign those two signals to the attested eruption of Etna,[28] it was necessary to do so with reservation. The trouble is that the concentrations of those peaks appear to be too massive to fit the presumed size of Etna's eruption in 44, even if we allow for the sulfur-rich content of Etna's gases.[29] They imply such a violent event that the impact on Roman society should have been much greater than our sources attest, and we would expect to find in the geological record unmistakable signs of such a powerful eruption.[30]

This difficulty now disappears thanks to the recent discovery of a much smaller acid peak dated to 43±2 B.C. in a third ice core, GISP2 (Greenland Ice Sheet Project 2).[31] This signal is separate from a massive one in the GISP2 which falls between the years 54 and 51 B.C. and is most likely identical with the one found previously in the Camp Century and Dye 3 ice cores. Zielinski speculates that the signal peaking in 53 B.C. could well have been produced by the eruption of an Icelandic volcano, whose proximity to Greenland would account for the remarkably high acid content. It is, after all, the second largest signal over the nearly 2100-year period from 100 B.C. to A.D. 1985 (Zielinski, Fig. 1a), and it dwarfs the size of the acid peak that is identified by Zielinski with the considerable eruption of Vesuvius in A.D.

---

[27] Hammer 56 (Fig. 2) compared the signal in the Dye 3 core, which he dated to ca. 50 B.C., with the signal in the Camp Century core and found the latter to be 2.3 times larger.

[28] E. g., Forsythe 49 n. 3, Bicknell 3 n. 7.

[29] The signal in the Camp Century core for 50±30 B.C. (192 kg km^{-2}) is more than three times the size of the one produced by the violent eruption of Tambora in 1815 in the Crête ice core (58 kg km^{-2}): Hammer et al. 231, Table 1. It is true, however, that Etna is a rich source of SO_2, accounting for close to 10% of the volcanic output of SO_2 globally: Allard 388; cf. Haulet et al. 715 and Jaeschke et al. 7253.

[30] So Hammer et al. 233, who conclude (repeated four years later by Hammer 56) that the source of the signal in 50±30 B.C. is "unknown". Stothers et al. 6360, while acknowledging the difficulties of connecting the major signal ca. 50 B.C. with the eruption of Etna in 44, nonetheless viewed Etna as the best candidate.

[31] Zielinski et al. 950 and in greater detail, Zielinski 20,947-48. We thank Greg Zielinski for discussing his findings with us by e-mail and for sending us an advance copy of his 1995 article in *J. Geoph. Res.*

79.[32] By contrast, the signal in 43 B.C. (SO_4^{2-} residual of 30 ppb), approximately one-tenth the size of the one in 53 (291 ppb), could well have been produced by the eruption of Etna attested in our literary sources. Since it can take one to two years after an explosive eruption for the sulfates in the aerosol to be deposited in the ice,[33] and since the biyearly samples taken from the GISP2 ice core introduce a possible dating error of ±2 years, the sample assigned to 43 B.C. is well within the parameters that would be expected following an explosive eruption of Etna in 44.[34]

The continent of North America furnishes the fourth and final independent strand of evidence pointing to an explosive eruption of a volcano in the latitude of Mt. Etna in, or about, the year 44 B.C. Studies of the annual rings of the subalpine bristlecone pine from one site in the White Mountains in eastern California (37° 30'N), Campito Mountain, where the treeline chronology has been extended back to 3435 B.C., reveal frost damage in 42 B.C.[35] This record is now confirmed by evidence found at another upper timberline site in California, Cirque Peak, whose latitude (36° 25'N) is also within a few degrees of Etna's (37° 44'N) and Ch'ang-an's (34° 17'N). Samples taken from the temperature-sensitive foxtail pine have revealed a marked decrease in ringwidth between the years 44 and 43, and the attenuated growth persisted in 42 and 41 (Scuderi [2] 74-5). This decrease is equal to -2.0 SD (standard deviations) in the sensitivity values calculated to analyze the change in ringwidths, and it implies that the temperature of the growing season experienced a reduction of approximately 1° C (Scuderi [1] 113). Since a significant correlation has been shown to exist between decreases in ringwidths, which reflect lower growing season temperatures, and volcanic activity that has produced atmospheric veiling,[36] it is tempting to

---

[32] The SO_4^{2-} residual of the 53 B.C. signal is 291 ppb, as compared with 95 ppb for the signal in A.D. 78±2 (presumably Vesuvius): Zielinski, Tables 4 and 6.

[33] Hammer et al. 231; Zielinski et al. 948.

[34] The cumulative dating error for samples at that level of the core is a possible ±10 years (0.5%), but the calibration of the signal that is thought to be the eruption of Vesuvius in A.D. 79 (arrived at by counting the annual layers and found by this method to be within 10 years of A.D. 79) tends to assure the accuracy of the dates assigned to the peaks in 53 and 43 B.C.: Zielinski 20,940, 20,948.

[35] LaMarche-Hirschboeck 125 suggested that this event was possibly to be connected with the large acidic spike found by Hammer et al. in the Camp Century ice core in 50±30 B.C. and was most likely the result of the eruption of Etna in 44.

[36] Scuderi (2) 72-3 found significant reductions in ringwidths corresponding to 8 of the 13 major volcanic eruptions during the 1380-yr period from A.D. 600 to 1980 (p. 13 Table 1). Within the past 100 years, LaMarche-Hirschboeck 124 detected frost damage within two years of four major eruptions in the Northern Hemisphere or equatorial region, beginning in 1884 in response to the eruption of Krakatoa in 1883.

view the decreased size in the width of the tree rings between 44-42 B.C. as offering further confirmation of an eruption in the period when Etna is said to have been active.

Although it is impossible to be certain that an eruption of Etna lies behind all of this evidence pointing to volcanic activity in the late 40's B.C. (nor is it necessary for our purposes to prove that the responsible volcano *was* Etna), what we know of the distances and timing of these events fits Etna rather well. First of all, as previously noted, Etna is a sulfur-rich volcano, and recent studies have demonstrated that eruptions which give off large quantities of sulfur dioxide (SO_2) have the greatest climate-altering effect.[37] It is not the silicate micro-particles and ash that produce the greatest effect on the Earth's atmosphere, but rather the sulfur dioxide which, when it combines with water vapor and ambient oxygen, is converted into liquid droplets of sulfuric acid (H_2SO_4). Next, studies of the recent eruptions of El Chichón (1982) and Mt. Pinatubo (1991) have shown that the cloud of sulfur dioxide given off by a volcano typically circumnavigates the globe in two to three weeks,[38] and the length of time it takes for this sulfur cloud to be converted into the sulfuric acid aerosol that causes the cooling effect and unusual solar phenomena can vary from three to four weeks, or even longer (Rampino et al. 79). The density of this aerosol gradually increases until it peaks some months after the eruption,[39] and the mist can remain suspended in the stratosphere, 15 to 25 km. (approx. 9-15 mi) above the Earth, for months, or even years, after the initial eruption.[40] Finally, it has also been found that eruptions outside of the tropics have the greatest potential to alter the climate when they occur in the spring or fall since the air flow is greater at those two times of the year (AGU Report 2).

Livy's statement that Etna gave off a great quantity of flame before the Ides of March is in perfect harmony with the tradition in our other sources

---

[37] For instance, although the eruptions of Mt St Helens (1980) and El Chichón (1982) were both VEI 5 eruptions (Bluth et al. 329), the former gave off only a small quantity of SO_2 and produced only a half megaton of aerosols, whereas the sulfur-rich eruption of El Chichón produced 12 megatons of aerosols (AGU Report 15) and had a much greater climate-altering effect (Rampino et al. 81-2). We thank Bonnie Bright of the American Geophysical Union for sending us a colleague's personal copy of the 1992 Special Report since it could no longer be ordered from stock.

[38] The tracking of these sulfur emissions by satellite imagery is reported by Robock-Matson 195 and McCormick et al. 399. We thank Greg Zielinski for referring us to this literature.

[39] For instance, it took 9 to 12 weeks for the maximum density of Pinatubo's aerosol to be achieved: AGU Report 17.

[40] The mass loading of the stratosphere by the eruption of Mt Pinatubo decreased over a two and one half year period (McCormick et al. 401).

that the Sun was surrounded by a corona when Octavian entered Rome in early May. The sulfur dioxide given off by an eruption of Etna in late February, or early March, would have had quite enough time by May to have been converted into an aerosol whose density should have been nearing its peak. To judge from the "reddish" color of the comet, as described by the Chinese in May-June, the effect of this aerosol was already being experienced in China as well (Pang et al. 880). The timing is about right for an eruption of Etna prior to Caesar's death on the Ides of March to produce such a phenomenon in the skies over China in late May or early June; in 1815 it took approximately three months for the unusual optical effects of the Indonesian volcano Tambora to become noticeable in Europe (Rampino et al. 83). Since our Chinese sources, unlike their western counterparts, do not typically associate the color red with comets,[41] the description of the reddish cast of the comet in 44 B.C. most likely points to the existence of unusual atmospheric conditions. In one or two other instances where the Chinese sources state that a heavenly body appeared "red" (e.g., the comet of A.D. 178[42] and the supernova of A.D. 1054[43]), there is reason to suspect that the reddish cast may have been caused by a volcanic dust veil.

If the source of the veil in 44 B.C. was Etna, and if Etna experienced a series of eruptions during the course of the year, commencing with the one before the Ides of March, then the skies over Italy could have been even more occluded in May and June than those over Asia. For this reason, the comet of 44 may have been less visible to observers in Italy than it was to the imperial

---

[41] In western literature, "bloody" is a stock description of comets, which are said to "glow blood-red" (*cometae sanguinei . . . rubent*, Verg. *Aen.* 10.272-73), have "bloody hair" (*crine sanguineo*, Pliny, *NH* 2.89), or bathe the sky with "bloody fire" (*igne cruento*, Calp. Sic. 1.80).

[42] Ho no. 106, the only other "red" comet through the year A.D. 200 in the sources translated by Ho. During the reign of the emperor Ling Ti (A.D. 168-189), unusual solar phenomena of the type observed in 44 B.C. are reported, and Wilson et al. (1980) have suggested a causal connection with an eruption of the New Zealand volcano Taupo, which they date ca. A.D. 186. However, the date of this eruption, which Simkin-Siebert 54 give as A.D. 180?, may be somewhat earlier still as indicated by the ruddy appearance of the comet of 178 and by the discovery in the Dye 3 ice core from Greenland of what appears to be a tephra layer dated to ca. A. D. 174-175 (Hammer 53).

[43] The supernova of 1054, which may not have been seen from Europe because of a dust veil (see p. 99.13), is described by the Chinese as taking on a reddish hue in the final days of its visibility. A possible parallel for these reports from China may be found in the description of the comet of 1556, as seen from Nuremberg, which is said to have had a "reddish" head and "straw-colored tail" (Vsekhsvyatskii 105). Could it be that the unusual appearance of the comet of 1556 was influenced by the eruption of Haku-San in Japan, which commenced in April 1554 and lasted 88 weeks into 1556 (Simkin-Siebert 90)?

astronomers in China. The best evidence that there may have been more than one eruption of Etna in 44 is furnished by the reports of weather conditions from China (confirmed by the tree-ring record from North America)[44] and by the curious report in one of our western sources that a pitch blackness (characterized as a *solis defectus*) passed over Rome on May 14 and lasted "from the sixth hour until night" (*ab hora sexta usque ad noctem*, Servius on *Georg.* 1.466). Such a phenomenon could be explained as the product of an eruption of Etna towards the end of April. In the previous century, an eruption of Etna in 1886 (said to have been in some ways similar to the one in 44 B.C.) blocked out the Sun in Sicily, and it took two weeks for the dust veil to spread from Sicily throughout Italy.[45] Given all of this evidence pointing to a thick volcanic aerosol over the skies of Italy in the spring of 44 B.C., it should be quite easy to understand why casual observers in Italy, in contrast with the trained professionals in China, apparently failed to catch sight of Caesar's comet in May and June.

THE SIGHTING IN JULY

Turning next to the summer of 44 B.C., we find that neither the Chinese sources nor the letters of Cicero attest the comet that was seen during the course of Octavian's games in late July, and yet, as we have demonstrated in the previous chapter, there is good reason to believe that a comet was indeed visible. We should want to know, therefore, why two important and independent bodies of evidence fail to confirm the sighting in July. Let us take a closer look at this negative evidence from silence to see if it is grave enough to call into question our earlier conclusion that the comet was a historical reality.

Missing Chinese Report

We shall start with the Chinese sources and offer the following

---

[44] An explosive eruption in the first half of the year (Jan.-June) should have had a severe effect on the growing season of 44 (Scuderi [1] 111), and yet the weight of the evidence from China and the evidence furnished by the tree rings in North America point to 43 as the first year in which a significant alteration in climate occurred. The eruption that caused this change would most likely have occurred in the latter half of 44 (July-Dec.), and so Bicknell 9-10 speculates that the eruption reported by Livy may have been merely the first in a series.

[45] Stothers-Rampino 6360. The massive eruption of Tambora in 1815 produced up to two days of darkness as far away as 600 km. (Rampino et al. 83), which is approximately the distance between Rome and Etna (roughly 625 km.).

explanations, in ascending order of likelihood, to account for the failure of the July sighting of Caesar's comet to find a place in the extant Chinese sources.

(1) First, it could be that the comet was indeed seen from the Chinese capital in July but the officials in charge of the imperial observatory suppressed notice of it. Since the comet had earlier been drawn to the emperor's attention in May-June and had elicited an edict of contrition (quoted above, p. 80), it might have been considered impolitic to report the return of that comet, on a yet grander scale, some two months later. Admittedly, however, this possibility seems rather remote since every precaution appears to have been taken to insure that the records of unusual heavenly phenomena were as complete and error free as possible.[46] It is hard to imagine that the Confucian officials, who were so powerful at the court of Emperor Yüan (see p. 80.52), would run the risk of suppressing the report of such a potentially ominous omen as a daylight comet, if one had been observed.

(2) Next, if it is unlikely that a record of the sighting in July was deliberately suppressed, could it be that a record somehow failed to be made in the first place because circumstances prevented the Chinese from observing the comet in July? It is not difficult to imagine how such a failure could have occurred. Since the comet is said to have been visible from Rome for only seven days, the imperial astronomer in China could have missed observing it if the skies over Ch'ang-an happened to be cloudy for the brief span of a week.[47] The trouble with this hypothesis, however, is that the Chinese bureaucracy is known to have recorded reports of celestial phenomena that could not be observed from Ch'ang-an itself because weather conditions were unfavorable. The extant account of a solar eclipse assigned to 29 March 15 B.C. specifically informs us that the eclipse had been reported to the Astronomical Observatory from outside the capital and could not be observed at Ch'ang-an because of overcast skies (Dubs II 421.vii and III 552). Presumably, therefore, if a daylight comet was seen in late July, word of this sighting reached Ch'ang-an and found a place in the official archives even if the imperial astronomer could make no observations of his own because of the cloud cover.

---

[46] See Needham III 191 for the care taken to insure that false or mistaken reports of unusual phenomena would not be made.

[47] This explanation was suggested to us by David Weible, our colleague in German. Our consultant John Major informs us that it is not at all unusual in the present age for the sky to be overcast for a week or more during late July in modern Xi'an, which is situated near the site the Han capital, Ch'ang-an. See Dubs III 553, for a table giving the average number of overcast days, month by month, for the region of Ch'ang-an (based on the years 1924-1936). The average given for July just happens to be *seven days*.

(3) Finally, we must bear in mind that our extant sources clearly reflect but a small fraction of the records that were once kept by the imperial Astronomical Observatory, and the records that do survive are far less detailed and complete than the accounts from which they were drawn.[48] We should not, therefore, conclude from the absence of a report in the Chinese sources available to us today that the sighting of a given comet was not necessarily made from China, provided that the sighting is well documented in our western sources (as it is for 44 B.C.). If we analyze the East Asian sources collected and translated by Ho, we find that they preserve an account of a naked-eye comet about every five years, for an average of roughly twenty such comets per century (Stephenson [1990] 234).[49] However, based upon data from the past few centuries, for which our records are more complete, it seems that something in excess of 80 naked-eye comets may become visible each century (see p. 73). Many of those comets would be extremely faint objects, to be sure, (magnitude = +5 or +6), but since we can tell that quite a few of the extant records in the Chinese sources most probably concern such comets (see p. 97), it is reasonable to conclude that the trained sky-watchers in China would have detected and taken note of the vast majority of comets that attained naked-eye visibility.

Next, we can form some impression of how detailed the imperial archives must have been both from a silk manuscript that was recently discovered in a Han tomb dating to 168 B.C. (Hunger et al. 45) and from the lengthier accounts of certain comets that just happen to be preserved in our extant sources.[50] The silk manuscript provides relatively detailed accounts of the heliacal risings and settings of various planets during the years 246 to 177 B.C., compared with which there is nothing on the same scale in the extant Chinese reports of comets.[51] The reports of comets that we do have are found in the various Chinese dynastic histories, and the chance survival

---

[48] The records in the imperial archives no doubt perished in the turmoil that followed the overthrow of the usurper Wang Mang, either in A.D. 23 when the imperial palace was burned, or in A.D. 25 when Ch'ang-an itself was sacked by marauding bands (Dubs III 124).

[49] In keeping with this average, there are 18 cometary records for the first century B.C. (Ho nos. 47-64).

[50] E. g., Ho no. 32 gives an almost day-by-day account of a comet that first became visible on 6 August 147 B.C. and disappeared from sight after ten days on the 16th. Likewise, Ho no. 61 gives a very full account of Comet Halley in 12 B.C. from 12 August until it was last seen 56 days later.

[51] The fragmentary astronomical records on clay tablets from Babylonia, taken with the silk MS from China, permit us to surmise how detailed the Chinese imperial archives may have been. For comets, the Babylonian tablets typically report the first sighting, heliacal setting and rising, stationary points and last sighting (Stephenson [1990] 244).

and fullness of those records was quite clearly dependent upon the nature of the intermediary sources upon which the authors of the Chinese dynastic histories drew. As the histogram constructed by Stephenson (1990) 239 (Fig. 13.3) for the years 209 to 1 B.C. reveals, there seem to have been at least two of these secondary sources available to the authors of the *HS* in the latter half of the first century A.D., and one of those sources was fuller than the other.[52] That same histogram shows that for the period 60-30 B.C., in which the comet of 44 falls, the records found in the intermediary source(s) were apparently sparse. Therefore, the Chinese account of the comet of 44, which tells us only how that comet appeared in the fourth month (18 May-16 June), is best viewed as but a fragment of the fuller account that is likely to have existed at one time in the imperial archives. The more complete records in the imperial archives could well have contained entries for each of the seven days in July during which the comet is reported to have been seen from Rome.

At first glance, it may appear that a fragment of the missing Chinese record for late July found its way into a Korean text translated by Ho (147). The author of that text drew upon Chinese sources for astronomical data,[53] and although the date ("fourth month") and location ("*Shen*") of the comet as given by the Korean text match the information found in the "Treatise on Astronomy" (*HS* 26: quoted above, p. 75), the comet is described differently. (Pointed and square brackets are employed, as before, for supplements.)

> "<During the> summer, <in the> fourth month <of the> fourteen<th> year [of King Pak Hyokkose, legendary founder of kingdom of Silla in 57 B.C.] there was <a> star sparkling (*hsing-po*) in *Shen*."
>
> (*Chronicle of Silla* in the *Samguk Sagi* 1/2).

Since the term *hsing-po* (literally "star bushing out") came at an early date to denote a comet that shoots out its rays in all directions,[54] and since

---

[52] Direct access to the imperial archives was, of course, by that date impossible: see p. 109.48.

[53] Kim Pu-sik, the Confucian scholar, who completed in A.D. 1145 the partly mythical *Samguk Sagi* ("History of the Three Kingdoms", from antiquity through A.D. 936), undoubtedly drew upon Chinese sources for the earliest period since historical records in Korea probably did not predate the 4th cent. A.D. Borrowing from the *HS* is demonstrated by the false report of an eclipse on 23 Aug. 34 B.C., which is found in both the *HS* and *Samguk Sagi*: see Gardiner 46, 65. We thank Prof. Sarah Nelson of the University of Denver for discussing with us the nature of these Korean texts and their probable sources.

[54] So defined, in contrast with a *hui-hsing* (lit. "broom star": i.e., a comet having a discernible tail), by the early third-century A.D. commentator Wen Ying (ad *HS* 4:13b), cited by Loewe 9. See also Ho 136, quoting from the Chin-shu 14:4a (composed in the

several of our Western sources describe the comet in July as giving off rays in
all directions (Texts 6, 13, cf. 3), it is tempting to see in the Korean text a
possible allusion to a Chinese sighting in July. The reference to a *po* comet
could conceivably have been transferred from July to May-June if the author of
the *Chronicle* carelessly condensed a more extended account of the sort that
is preserved, for instance, for the year 119 B.C. when, according to the *HS*
(6:16a), a *hsing-po* ("a sparkling comet") was seen in the spring and a
ch'ang-hsing ("long comet") was seen in the summer. Upon closer
inspection, however, we find that this notion must be discarded. No special
significance can be read into the description of the comet of 44 as a *po* comet
in the Korean *Chronicle* because the source upon which the Korean author
drew is demonstrably a passage in the *HS* "Annals" (9:5b)—a text not cited
by Ho or Zhuang-Wang. The two texts are identical, apart from the
conversion of the Chinese reign period to the equivalent year of the Korean
King Pak Hyokkose, and it can be shown (e.g., Loewe 7-9) that in the *HS*
"Annals" the expression *hsing-po* is a generic term for a comet, without any
reference to how the comet appeared physically, whether with or without a
visible tail. All that remains, therefore, in our Chinese sources is the heavily
abbreviated account of the sighting in May-June 44 B.C.

Before leaving this topic, it is instructive to examine the process of
selection that most likely caused an account of the earlier sighting in
May-June to be preserved, while the later sighting in July, which figures so
prominently in our Western sources, has left no trace in the extant Chinese
sources. Significantly, we learn from *HS* "Annals" that the sighting in
May-June assumed some historical significance because it inspired Emperor
Yüan to issue an edict (quoted above, p. 80). In the West, by contrast, the
sighting in May-June fell through the cracks presumably because there was
no comparable event in Roman history to insure that the comet would become
part of the historical tradition. In July, just the opposite was true: the Chinese
historians failed to link the comet with anything of moment, whereas in Rome
the chance appearance of the comet during Octavian's games (and the role it
played in securing the recognition of Caesar as a god) insured that it would
find a place in the historical tradition. We suspect that such factors as these
played a very significant role in determining which reports of comets were to
survive from antiquity. A select number of comets (and sightings of comets)
are known to us because historians chose to include them in their accounts of
contemporaneous historical events. Those comets are preserved like so many
flies in amber thanks to their historical contexts, while the vast majority of

early seventh century A.D.): "By definition a comet pointing towards one preferential
direction is a *hui*, and one that sends out its rays evenly in all directions is a *po*."

comets have come and gone without leaving a trace simply because they did not happen to coincide with events of historical moment.

A few further examples such suffice to establish this principle of selection. As a control, we shall confine our examples to Comet Halley, whose historical reality is not open to doubt. The Chinese sighting of 1P/Halley in 240 B.C. was assured a place in our extant histories, at least in part, because it happened to occur at the time of General Ao's death and a few months prior to the death of the Empress Dowager (Stephenson [1990] 250). Those two events attracted the notice of historians, and when they sifted the historical records for the period, they found and passed on to us the account of 1P/Halley. In like fashion, the return of Comet Halley in 87 B.C. was preserved in our Western sources because it chanced to fall in the year in which Cinna and Marius captured Rome; significantly, it is the only historical comet of which Cicero took any notice (*Nat. D.* 2.14). On its next return, in 12 B.C., the same comet is linked in one of our Chinese sources with an edict issued by the emperor (Stephenson [1990] 234), and in the West it was remembered because it is included among the portents that attended the death of Agrippa (Dio 54.29.8). And finally, the return of 1P/Halley in A.D. 1066, which is commemorated in the Bayeux Tapestry,[55] came inevitably to be regarded as an omen foretelling the overthrow of King Harold of England in the Norman conquest.

These examples could be multiplied manyfold, but those given should suffice to demonstrate that comets were not always "attracted" from their correct date to the date of some great historical event. In fact, what may appear to be a suspicious coincidence may not be that at all. In the absence of some significant historical event, it is unlikely that the knowledge of a comet, an earthquake, an eruption of a volcano, or the like, will be handed down to later generations. Such events will, of course, be vividly remembered by those who experienced them (doubtless associated in the popular mind with some happening of local significance such as the death of a local nabob, a poor harvest, and so forth), but an account of comets and the like will not become part of the historical record unless there happens to be some historical peg on which the comet can be hung.

Silence of Cicero

Let us next address the silence of Cicero's letters concerning the sighting of the "Julian star" in July. The failure of Cicero to mention the comet may in part simply reflect the failure of the comet to impress itself upon each and every contemporary observer. Although, according to Augustus in his *Memoirs*, the

---

[55] Illustrated in Sagan-Druyan 27.

comet was "bright and plainly seen from all lands" (Text 1), we should not be surprised if the grandeur of the event was magnified by Augustus and his supporters to the greatest extent possible.[56] The influence of that tradition doubtless colored later accounts, such as Obsequens' assertion that the comet "caught the eye of everyone" (*convertit omnium oculos*, Text 4). Such remarks are best taken with a grain of salt.

As a matter of fact, daylight comets are usually not terribly easy to see, unless the observer knows where to look. For instance, the so-called Great Daylight Comet of 1910 (1910 I, not to be confused with 1P/Halley 1910 II) did not cause much of a stir among the public because the average person did not know where to look for it (Richardson [2] 205). Furthermore, the comet of 44 B.C. may not have been visible during the daylight hours from Cicero's vantage point. We know that he was making a voyage to Greece and traveling along the coast of southern Italy from 17 July to 1 August. Apart from two brief stopovers, at Velia (19-20 July) and Vibo (24-25 July), Cicero appears to have spent the remainder of those days on shipboard during the day, and probably on shore during the night, since the ships on which he was traveling were small and not likely to have remained at sea after dark.[57] Presumably Cicero's little flotilla made land before sundown, and Cicero put up for the night in the shadow of the Apennines. In that region there is generally a narrow coastal plain of less than one kilometer, and at the edge of the plain the mountains rise from an initial elevation ranging from 50 to 200 meters up to 1000 meters farther inland.[58] Therefore, if we were correct in drawing the conclusion in the previous chapter that the *sidus Iulium* rose in the NNE, Cicero most likely did not see it until well past sunset because the

---

[56] George Goold puts this point nicely in written communication (27 March 1994): "The silence of Cicero scarcely dispels the suspicion that Augustus (Octavian) magnified everything to do with the comet so that it could give credence to the story he desired to spread." Cf. above, p. 63, and see Appendix V for a tabular list of the parallels revealing the probable influence of the *Memoirs* on our other sources.

[57] There were three ships in Cicero's flotilla (*Att.* 16.3.6), and they were small, as we can tell both from the statement that they were only "ten-oared" and from their being called *actuariolae*, the diminutive of *actuaria*, a type of light warship used for transport: see Leubeck, *RE* 1 (1893) 331 and Dar.-Sag. I.1 59-60. Typically ships of war, as opposed to freighters, which were the usual way for passengers to travel, had to put in to harbor by night: see Casson 151-54. For the diminutive *actuariola*, which Shackleton Bailey aptly translates as "rowing-boat", cf. *Att.* 10.11.4, where it is plain that a small craft is to be envisaged. One reason why Cicero may have chosen these small, maneuverable military craft, in preference to a freighter, was his fear of piracy off the south coast of Italy. He had hoped to travel under the protection of Brutus' fleet, but Brutus showed no signs of making a timely departure (*Att.* 16.4.4; 2.4).

[58] See U. S. Army Map Service, *Italy 1:250,000* (Washington, D.C., 1944), sheets 42 and 47.

mountains will have blocked his view of the horizon in the NNE. If he saw it at all, he did so when the sky was filled with stars, and despite its brilliance, since the "Julian star" had little, or no visible tail (see p. 139), it may not have excited much interest on Cicero's part or the part of locals in southern Italy where he was camping. Still, it may seem surprising that Cicero took no notice of the comet if only for the reason that it played such a prominent role in winning public acclaim for Octavian's games and contributed momentum to the movement aimed at securing the recognition of Julius Caesar as a god.

As chance would have it, however, one and only one of the extant letters from July and August 44 *could* have made mention of Octavian's games and so given us Cicero's reaction to the news of how Octavian greeted the comet as a sign of Caesar's apotheosis by placing a star on the statue of Caesar.[59] Only five or six letters survive from the period beginning with Cicero's departure from Pompeii for Greece on 17 July (three days before the commencement of Octavian's games during which the comet was seen) and ending with his return to Rome on 31 August: *Att*. 16.3, written on 17 July; *Att*. 16.6 of 25 July, written at Vibo in southern Italy; *Fam*. 7.20 and 19 of 20 and 28 July to Trebatius from Velia and Regium in S. Italy; *Fam*. 11.29 written sometime in July to Oppius; and finally *Att*. 16.7 of 19 Aug., written on the return voyage to Pompeii. Only the last of these letters (*Att*. 16.7) could have contained Cicero's reaction to the news of Octavian's games and how Octavian exploited the appearance of the comet; in the previous letters Cicero was completely out of touch with Atticus and news from Rome. This gap in Cicero's communication with Rome and Atticus lasted from the date of his departure from Pompeii (17 July) until sometime after 7 August. We can be sure of this fact (1) from his account in *Att*. 16.7.1 of the startling and (as it turned out) stale news from Rome that Cicero learned on 7 August from some citizens of Regium (on the toe of Italy) who had recently returned from a visit to the capital and called on Cicero, who was staying near their town;[60] and (2) from Cicero's discovery shortly after his interview with the

---

[59] This fact remains true whether or not we follow SB in interpreting *reliquorum ludorum* of *Att*. 15.26.1 as a reference to Octavian's games about which Cicero wished to keep him informed.

[60] These visitors to Rome may well have left the capital before the commencement of Octavian's games on 20 July and so have had no knowledge of the reaction to the comet in Rome. Their departure certainly preceded 1 Aug. because they knew that the Senate was to meet on that date and yet they did not know the outcome of that meeting. We can determine the likely day of their departure more precisely by working back from 7 Aug., the date of their interview with Cicero (*Att*. 16.7.1, combined with *Phil*. 1.8). Since 7 days is the record time for a journey between Rome and the Strait of Messina (*Att*. 2.1.5), the Regians had to leave Rome by 31 July to arrive on 6 August, the day before they called on Cicero; if they reached Regium a few days earlier, as is likely, then their departure from

Regians that Atticus had radically changed his mind about the appropriateness of
Cicero's intended voyage to Greece.[61]

Given the chance survival of these particular letters and the gap of more than
two months that separates them from the next in the series to Atticus (*Att.* 15.13
of 25 Oct.),[62] it should come as no surprise that Cicero fails to mention Caesar's
comet. By October, when the correspondence with Atticus resumes, the comet and
games of July were decidedly old news; at that time, Mark Antony appeared
poised to crush all rivals, including Octavian, by taking charge of the legions that
had arrived at Brundisium from Macedonia. Meanwhile, Octavian was busy raising
troops composed of Caesar's veterans, and by early November Octavian was
sending letters daily to Cicero, urging him to save the Commonwealth a second
time (*Att.* 16.11.6). If we look more carefully at the one letter written in August
that could have discussed the comet in the context of news from Rome after Cicero
re-established contact with Atticus, it is perfectly clear why the comet and games
find no place in that letter (*Att.* 16.7 of 19 Aug.). The letter has but one theme: it is
wholly concerned with thrashing out the merits of Cicero's decision to sail for
Greece and how his journey was perceived by his friends and the Roman public.
Every word dwells on that one subject, and there is no space allotted to chitchat or
matters unconnected with Cicero's aborted voyage to Greece.

Finally, it is instructive to recall that the comet is by no means the only
event during these same weeks to escape mention in the *preserved*
correspondence of Cicero and his friends. One example of another such "nine
days' wonder" should suffice to illustrate this point: Shortly after the games
(i.e., in late July, or early August) Mark Antony and Octavian are said to have
participated in a short-lived, public reconciliation of their feud. This reconciliation
was staged in a ceremony on the Capitoline, at the urging of Caesar's veterans
and with their strong approval (see our p. 2). It raised high hopes on the

---

Rome must have preceded 31 July by an equal number of days. Finally, since there is no
indication that Regians travelled rapidly, and since we know that despite his avowed
haste (*Phil.* 1.9; *Fam.* 12.25.3), Cicero required 9 or 10 days to journey from Regium to
Velia (8-17 Aug., Att. 16.7.5), less than half the distance to Rome, we can easily put the
departure of the Regians slightly before the beginning of Octavian's games on 20 July.

[61] We can tell that Atticus' letter caught up with Cicero sometime after he met with
the Regians because Cicero states that as a result of that interview he had already made
up his mind to return to Rome when he subsequently read Atticus' criticism of his intended
voyage to Greece (*Att.* 16.7.2).

[62] None of the letters *Ad fam.* in Sept. offers an appropriate context for a discussion
of the games and comet; *Fam.* 11.27 to Matius, which SB assigns tentatively to mid
October (see intro. n.), does indeed touch on Octavian's games (Text 16), but since Cicero
wished to put the best face on Matius' role in producing the games we can hardly expect
Cicero to dwell upon details that he must have found distasteful.

part of the public that the two competing factions among Caesar's followers would draw together and present a united front against the anti-Caesarian, aristocratic faction in the Senate, and yet nowhere does Cicero allude to this event. Nonetheless, scholars find no reason to distrust the substance of the report in our other sources simply on the grounds of Cicero's silence, and we should likewise be wary of assigning undue importance to Cicero's failure to remark upon Caesar's comet. As we have seen, this silence is most easily explained in part by Cicero's isolation from news of events that were unfolding in Rome in late July, and in part by his overriding concern, when news from the capital finally caught up with him, to respond to the criticism that was being levelled against his departure from Italy. Later in the year, when Cicero was back in Rome and found himself increasingly drawn into a clash with Mark Antony, Octavian's games in July and the comet were decidedly news that was passé. On top of this, Cicero more and more wished to put the best face on Octavian's activities in the hope of using him as a potent ally in the struggle against Antony. This goal, of course, ruled out any possible reference to the comet in Cicero's *Philippics,* speeches that attacked Antony and emphasized the pro-senatorial leanings of Octavian. Obviously it would have been inappropriate in the *Philippics* to mention Octavian's agitation in July, in response to the comet, to have Julius Caesar recognized as a god.

Probability of the Two Sightings Being Attested
by Both Western and Far Eastern Sources

We must accept the fact, therefore, that our sources preserve accounts of the sighting of Caesar's comet on two separate and distinct occasions: by the Chinese in May-June and by the Romans in late July. We might wish for these records to overlap and confirm one another, but if we take into the consideration the extant records of comet sightings overall, we find that it is, in fact, statistically unlikely that reports of a given sighting will come down to us from both the Romans and the Chinese. Therefore, we should not be surprised to find that our two bodies of evidence for the comet of 44 concern different periods, rather than a single sighting. To take the first century B.C. as an example, Hasegawa lists 34 naked-eye comets seen during those hundred years, out of which 16 were seen by only the Romans and/or Greeks, 15 by only the Chinese and/or Koreans, and 3 by observers both in the Mediterranean and in the Far East. However, based upon the statistics in Hasegawa's catalogue for recent centuries, for which our records are fullest, there should have been as many as 87 naked-eye comets visible during the first century B.C. (see p. 73). The efficiency of our first-century B.C. Chinese "comet detector" was, therefore,

$$Pc = (15 + 3)/87 = 0.207$$

whereas the efficiency of the Roman detector was

$$Pr = (16 + 3)/87 = 0.218.$$

These "detection efficiencies" are, of course, actually the probability that records of a cometary observation by either of these two groups should have reached us. Based upon these figures, we can now estimate the number of comets that should have been recorded as seen by both groups:

$$Nboth = PcPr \times 87 \pm \sqrt{(PcPr\ 87)} = 4 \pm 2,$$

comparing well with the three attested instances in the first century. The probability that we should have records of the Chinese sighting in May, but not the Roman, and of the Roman in July, but not the Chinese, is

$$Pcr'c'r = Pc(1\text{-}Pr)Pr(1\text{-}Pc) = 0.0280,$$

and the probability that we should have records of both sightings by both groups is

$$Pcrcr = P_c^2 P_r^2 = 0.0020.$$

Thus, the combination of records that we actually have (Chinese for May-June, but not Roman; Roman for July, but not Chinese) is 14 times as likely to have happened as the combination that we would like to have.

THE PROBABLE ORBIT OF COMET CAESAR

The July Outburst

Abstract

We conclude from the silence of our sources for the period between the sighting by the Chinese (18 May-16 June) and the sighting by the Romans (ca. 20-23 July) that the comet was relatively dim before it underwent a sudden increase in luminosity in July. Comets are known to experience such flare-ups unexpectedly and at various distances from the Sun—in this instance almost two months after the likely date of perihelion. Such flare-ups have been recorded with an increase of up to nine magnitudes in luminosity. Two distinctive patterns of luminosity increase have been reported (illustrated in Figs. 5a & 5b). Assuming that when the comet was seen at the 11th hour (more than an hour before sunset) in July it had a brightness equal to at least that of planet Venus, which is visible in the daytime at magnitude -4, we obtain a lower estimate for the brightness of the comet prior to its sudden burst of brightness in late July. The range is shown to be between +1 and +5, depending upon which of the two curves for increases in luminosity we assume for the comet of 44 (Figs. 6a & 6b). If its brightness resembled that shown in Figure 6a, it would have been rather dim (magnitude of ca. +4 to +5) during the latter part of June and the first three weeks of July, requiring relatively sharp eyesight to be seen in the weeks leading up to the flare-up ca. 20 July. If, on the other hand, its brightness resembled the curve shown in Figure 6b, it should have been visible throughout most of June and during the first three weeks of July, perhaps resembling a bright star in the very late evening or early morning sky.

The weight of our evidence suggests that the comet was relatively dim between the initial sighting in May-June and its sudden emergence as a daylight-visible phenomenon in July. Such increases in comet luminosity are not unusual, can occur at almost any distance from the Sun, and have been known to take place more than a month after perihelion. The "Great Comet" of 1811 (1811 I), for instance, like the comet of 44 B.C., achieved its maximum brightness well over a month after its perihelion. That comet had an estimated magnitude slightly brighter than +4 when it passed perihelion on 12 September, but some forty days later it approached a magnitude of 0.[1] The comet of 1811, however, grew brighter because it was drawing closer to the Earth, whereas the comet of 44 B.C. seems not to have been appreciably closer to the Earth when it experienced a sudden, anomalous brightening caused by an outburst.[2] The maximum observed increase in the brightness of a comet experiencing such an outburst seems to be nine magnitudes at most. The mechanism is some sort of change in the internal structure of the comet's nucleus (Hughes). For a few comets the brightness increase is associated with a splitting of the nucleus (Sekanina 276-78). Recently the explosive polymerization of HCN has been discussed as a possible mechanism (Rettig).

A typical light curve for a brightness increase, as given by Richter ([1963] 143), is shown in Figure 5a. There is a rise in brightness over a very short time, then a period of roughly constant brightness, followed by a period in which the brightness falls to its initial value.

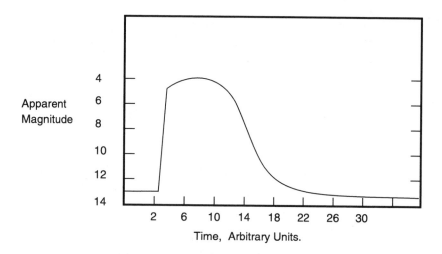

Figure 5a: Typical cometary brightness outburst (after Richter). The duration of the outburst varies, and is given here in arbitrary units. We have set the initial magnitude equal to +13, and the magnitude change equal to 9 in order to facilitate comparision with the curve in Figure 5b. The luminosity curve shown here has a sharp rise followed by a period of roughly constant brightness.

---

[1] Kronk 27 and personal communication (15 May 1996). We thank Brian Marsden for directing our attention to this comet in personal communication (23 Aug. 1995).

[2] On the nature of this outburst see p. 72, and see Figure 8 for the estimated distances between the comet and Earth both during and prior to the brightening.

Figure 5b shows the somewhat different light curve measured for the sudden outburst of comet P/Tuttle-Giacobini-Kresák on 7 July 1973 (Kresák), which had one of the largest magnitude changes on record. This comet showed an initial rapid increase in luminosity, rising to a peak in about four days, a steep fall-off over another four days to a level of moderate luminosity, which lasted for a few weeks, then a decline over a few days to its initial low luminosity.[3]

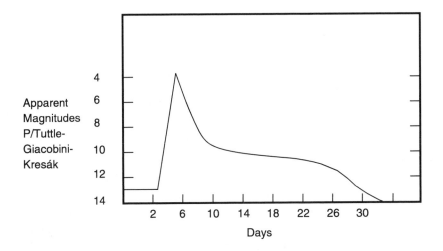

Figure 5b: Brightness outburst of comet P/Tuttle-Giacobini-Kresák, July 1973, (after Kresák). This comet's luminosity curve shows a very sharp peak. The brightness rose 9 magnitudes from day 2 to day 6, then fell 6 magnitudes over the next 4 days. Its luminosity declined gradually until day 26, then fell rapidly over the next few days.

The planet Venus is visible in the daytime at magnitude -4. Comet Caesar during its outburst must, therefore, have been at least that bright for seven days if we may trust the ancient sources (see p. 85). We can get a lower estimate of Comet Caesar's brightness just before the outburst by assuming that the change in brightness equalled the maximum recorded change of nine magnitudes. For a flat-topped luminosity curve like that of Figure 5a, Comet Caesar could have reached magnitude -4 if it started at

---

[3] See Yeomans 347 for photographs showing this comet, first on 4 July 1973 as a barely perceptible fuzzy light and then later, several days after the peak of the flare, as a brilliant object surrounded by an unmistakable coma.

magnitude +5. For the sharp-peaked curve of Figure 5b, Comet Caesar would have to start at something like magnitude +1 to have maintained magnitude -4 or better for a week. At its maximum brightness it would have been near magnitude -8 for a few days.

It is possible to estimate the comet's apparent magnitude **m** during the period before the outburst. Let H_1 denote the comet's absolute magnitude, Δ its distance from the Earth, **r** the comet's distance from the Sun, then (Marsden-Roemer 727)

$$m = H_1 + 5 \log(\Delta) + 2.5 \, n \log(r).$$

The distances Δ and **r** are determined by the comet's orbital position. The parameters H_1 and **n** depend on the comet's composition and surface conditions. The radial parameter **n** affects the rate at which the comet's brightness changes with its distance from the Sun. This parameter varies considerably from comet to comet, values between -2 and 11 being reported, with the most probable values between 3 and 6 (Richter [1963] 108). We do not know **n** or H_1 for Comet Caesar. They are however constrained by our estimates for the comet's brightness just before the outburst. Also, the Chinese report is not consistent with daytime visibility during May. The luminosity during May should, therefore, have been dimmer than magnitude -4, which puts another constraint on **n** and H_1. The range of possible light variations is illustrated in Figures 6a and 6b.

In Figure 6a, we model the July outburst after Richter's curve. We therefore estimate the apparent magnitude in July, immediately before the outburst, as +5. Taking **n** = 4 and H_1 = 3.3 gives an apparent magnitude of -3 in May-June. Between 10 June and 20 July its magnitude goes from about +1 to +5. It should have been visible during this period, though perhaps requiring sharp eyesight during the first three weeks of July.

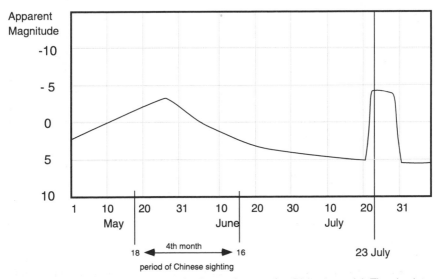

Figure 6a: A hypothetical light curve for Comet Caesar after Richter's model. The absolute magnitude and radial parameter are respectively set at 3.3 and 4. The July outburst is taken as a rise of nine magnitudes over a period lasting seven days. The comet's magnitude would have varied from about -2 to +5 during June-July.

In Figure 6b, we model the July outburst after that of comet P/Tuttle-Giacobini-Kresák. We estimate the apparent magnitude in July, immediately before the outburst, as +1. Taking $n = 3$ and $H_1 = 0$, we get an estimated apparent magnitude near -4 during late May, which is rather bright, but at that time it was close to the Sun and may not have been easily visible during the daytime. Between 10 June and 20 July its magnitude went from about -2 to +1, and the comet should have been visible, perhaps resembling a star.

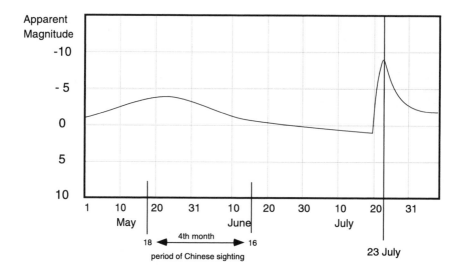

Figure 6b: A hypothetical light curve for Comet Caesar after that of comet P/Tuttle-Giacobini-Kresák. The absolute magnitude and the radial parameter are respectively set at 0 and 3. The comet would have been quite bright, magnitude about -4, near perihelion in May, though perhaps it was difficult to see because of its proximity to the Sun. It would have thereafter appeared at night as a bright star until its outburst in July, when it would have become visible during daylight.

The Orbital Parameters

Abstract

Typical orbits are calculated by starting from a point within each of the allowed regions for the attested sightings (from the Chinese capital in May-June and from Rome in July). We assign the date of 30 May to the first sighting and 23 July to the other. We take a perihelion distance of less than about 1/4 A.U. (which is a likely inference from the Chinese report) and fit the positions with parabolic orbits, both direct and retrograde, that is, rotating around the Sun either in the same or opposite sense as the planets. A Monte Carlo calculation is performed to determine the uncertainty in the orbital parameters. The parameters for two typical orbits are given in Table 1. The variation in comet behavior near the Sun is displayed in Figure 7a, where several orbits are plotted in right ascension and declination relative to the Sun. For one of the possible perihelion distances (0.224 A.U.) we compare in Figure 7b the behavior of the direct and the retrograde orbits near the Sun, and in Figure 8 we show the positions of comets having these orbits relative to the Sun and Earth on 30 May and 23 July.

An orbit requires the specification of six parameters (those given in Table 1, p. 126, plus eccentricity), and if we had a reliable account of three observations, rather than just the two sketchy reports from China and Rome, we could calculate the six parameters of Comet Caesar from the right ascension and declination of the three sightings. Unfortunately we lack precise accounts of the two observations that were made, and we do not have the crucial third point of reference needed to approach the solution from that direction. There are, therefore, an infinite number of orbits that will fit the two observations. On the other hand, the range of possible orbits for Comet Caesar can be narrowed considerably by taking into account two likely inferences that may be drawn from what we *are* told about the comet. First, we expect the comet's approach to the Sun to be through the cone of invisibility and the skyglow, as shown in Figure 2 (p. 82). This implies a perihelion distance of less than roughly one-quarter A.U. Second, the vast majority of comets with such small perihelion distances have nearly parabolic orbits (see p. 84); that is, their eccentricities are extremely close to 1.0. Hence, we may take 1 as being the likely eccentricity of Caesar's comet.

We have written a computer algorithm which takes two positions, plus a value for the perihelion distance, and generates the parameters for two parabolic orbits, one of inclination less than 90 degrees, and one orbit of higher inclination, which can be greater than 90 degrees. The comet in the lower inclination orbit will rotate about the Sun in the same sense as the planets, in a direct orbit. An inclination greater than 90 degrees corresponds to rotation in the opposite sense, in a retrograde orbit. The algorithm (see Appendix VIII) was constructed using well-known relations from celestial mechanics (McCuskey and Fitzpatrick), plus standard numerical algorithms (Press et al.).

We assigned the Chinese sighting to 30 May and the Roman to 23 July, each date near the midpoint of the allowed time intervals. Next we chose two positions close to the middle of the allowed regions on the two dates (the position on 30 May being adjusted so as to put the comet on the same meridian as the Sun, as shown in Figure 2). The positions chosen were 4h 14.2m, 35.0d on 30 May (Epoch -43) and 23h 55m, 45.7d on 23 July at 12:30 GMT. Two orbits with a perihelion distance of 0.224 A.U.,[4] one retrograde and one direct, were found that fitted these positions and gave reasonable behavior near the sun, appearing above the line of visibility for a few days (see Table 1 below). Based upon some earlier estimates of ours concerning a range of perihelion times, Brian Marsden[5] has very kindly provided us with a set of orbits that fit two slightly different positions, with somewhat different

---

[4] An arbitrary value that resulted from previous estimates based upon the Gauss method (McCuskey 86-91).

[5] Personal communication (9 November 1996).

perihelion distances (see Figure 7a for an illustration of one of these). His orbits also satisfy the requirement of a short period of visibility near the Sun.

To get an idea of the uncertainty in our resulting orbital parameters, we performed a Monte Carlo calculation. A five-dimensional "box" was defined in such a way as to enclose the reference positions given above for our two dates of observation (30 May and 23 July) as well as our 0.224 A.U. perihelion distance. The computer then chose positions and perihelion distances at random within the box.[6] It found 41 sets of orbital parameters within this box, and computed their mean values and standard deviations. The mean values were not significantly different from those of our 0.224 A.U. retrograde and direct orbits, which we give, along with the standard deviations, in the following table.

TABLE 1 [7]

	Perihelion	Inclination	Ω	ω	DP
Retrograde	0.224 ± 0.048 A.U.	109.81± 20°	141.40 ± 30°	17.09 ± 17°	25.16 ± 1 d
Direct	0.224 ± 0.048 A.U.	45.91± 10°	178.88 ± 22°	6.91 ± 23°	25.26 ± 1 d

Of these two orbits, the direct causes the comet to approach the Earth more closely during the latter part of May (see Figure 8), and so the comet should have been approximately one magnitude brighter in May-June than the comet having a retrograde orbit.[8] The apparent absence of reported sightings during that period in 44 B.C. (see pp. 91-94) would tend to favor a dimmer comet, making it slightly more probable that Caesar's comet had the

---

[6] The box was defined as follows: ± 0.08 A.U. for the perihelion distance, since 0.14 A.U. gives probably as close and 0.3 A.U. probably as distant a passage as might be visible for the requisite few days; ± 5° for the 30 May right ascension, this being the estimated uncertainty in the position of the Shen; ± 15° for the May declension (the lower limit is right on top of the Sun, while the upper limit would have the comet visible for too long a time); and for the July sighting, ± 15° in right ascension and ± 6° in declination brackets most of the allowed region. One of the numerical algorithms also required a limit to be placed on the acceptable date of perihelion. This was taken to be between 22.5 and 27.5 May, since a preliminary run indicated no orbits with dates outside this interval.

[7] This table lists the perihelion distance, inclination, longitude of the ascending node (Ω) in Epoch -43.0, argument of perihelion (ω), and the day of perihelion (DP) in May 44 B.C.

[8] See pp. 121-24 for an estimate of the comet's apparent magnitude. The difference in magnitudes between comets having direct and retrograde orbits, situated at the same distance from the Sun, is $5\log(\Delta dir/\Delta retro)$. The ratio $\Delta dir/\Delta retro$ of the Earth-comet distances is about 0.8 in May-June, which makes the direct comet brighter by about 0.7 magnitude.

retrograde orbit. We therefore base the remainder of our analysis on a comet with a retrograde orbit.

In Figure 7a we show three typical orbits, two retrograde (one with a perihelion distance of 0.304 A.U. and our standard 0.224 A.U. orbit) and Dr. Marsden's 0.139 A.U. orbit. This last has an inclination very near 90°, and so it is neither retrograde nor direct. The 0.304 A.U. orbit spends rather too much time outside the skyglow, the 0.139 orbit perhaps too much time within the skyglow.

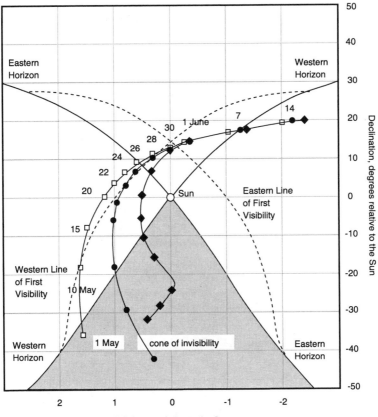

Figure 7a: Three hypothetical orbits for Comet Caesar. These orbits are plotted in right ascension and declination relative to the Sun. The western horizon at sunset and the eastern horizon at sunrise are shown. The lines of first visibility are drawn very schematically. The orbits have perihelion distances of 0.304 A.U., □, 0.224 A.U., ●, and 0.139 A.U., ◆. The apparent positions of the comets on these orbits almost exactly coincide on 30 May, and they follow nearly the same path across the sky for some time thereafter.

Figure 7b compares the behavior of our 0.224 A.U. retrograde and direct orbits near the Sun. They both spend about the same amount of time at the outer edge of the skyglow. The path of the direct comet against the sky from mid-May to July is not significantly different from that of the retrograde comet.

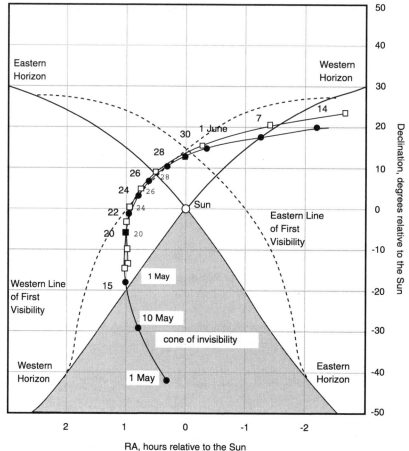

Figure 7b: Comparision of retrograde and direct orbits. Both orbits have perihelion distances of 0.224 A.U. and the apparent positions of the comets on these orbits coincide on 30 May and 23 July. The path across the sky for the direct orbit, □, differs only slightly from that of the retrograde orbit, ●.

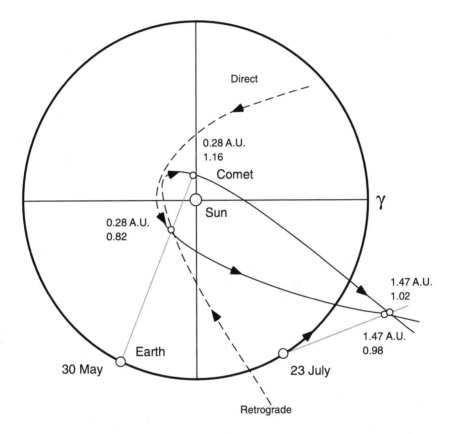

Figure 8: Two orbits for Comet Caesar, each with a perihelion distance equal to 0.224 A.U. One orbit is retrograde, the other is direct. The view is of the plane of the ecliptic from the north side. The portions of the orbits that pass beneath the plane are shown dashed. The Earth's orbit and its positions at the two dates of observation are shown, as well as the lines of sight to the comets following the orbits. The comets' apparent positions coincide on these dates. Next to each comet position, the Sun-comet distance is listed above the Earth-comet distance.

The Comet's Path as Seen from the Earth

Abstract

For the orbit of 0.224 A.U. we show in Figures 9 and 10 the rising and setting times from 1 May to 30 July as seen from the Chinese capital of Ch'ang-an (34° 17'N) and from Rome (41° 54'N). Figure 9 reveals that during 20 to 26 July the comet as viewed from Ch'ang-an would have risen one half hour at most before sunset. It would not, therefore, have equalled the spectacular daylight phenomenon that is reported in our western sources for the latitude of Rome. We furthermore point out that the setting times for the latitude of Rome (shown in Figure 10) may possibly account for a variant tradition in our Roman sources (attributed to Baebius Macer, Text 6) that the comet was visible at the 8th (as opposed to 11th) hour and that it was visible for three (as opposed to seven) days (Text 5). If the outburst in luminosity fit the pattern illustrated in Figure 6b, the comet may well have achieved a magnitude in the range of -7 for a brief span, and for that brief period of approximately three days it may have been visible to a sharp-eyed observer when it set in the NW close to the end of the 8th hour (e.g., at 14:34 on 23 July, the 8th hour ending at 14:31).

After the initial Chinese sighting, the comet could have been seen during the early to late morning hours, assuming that it still retained its initial brightness. In Figures 9 and 10, we show the rising and setting times for the Sun and our 0.224 A.U. comet, as well as the time of local noon, for the latitudes of the Chinese capital and of Rome respectively. The time is given in Local Mean Solar Time. Figure 10 also shows the beginnings of the 11th and 12th hours at Rome.

As seen from the Chinese capital during 20 to 26 July, the comet rose half an hour at most before sunset. It would not have been as impressive a daylight spectacle as it was in Rome, but it should have been visible later at night.

As seen from Rome, the comet rose a few hours before dawn during the first week in June. During 20-26 July, it rose within the 11th hour, and it set a few hours before it rose, as one would expect for a northern object. On 23 July, it set at 14:34. It is not said to have been visible in the forenoon, so presumably its brightness was only sufficient to enable it to be visible during the reduced sunlight of late afternoon. However, the 8th hour would have ended at 14:31 on 23 July, and if the comet was visible then to a sharp-eyed observer as it set in the NW, this might account for Baebius Macer's report that it was seen during the 8th hour (Text 6) and for Servius' statement (possibly based upon Baebius?) that the comet was visible "at midday" (*medio die*, Texts 6, 7). Furthermore, if Servius in Text 5 was also perhaps drawing upon Baebius, this might explain why Servius departs from the Augustan tradition and states that the comet was visible "for three days" (*per triduum*), instead of seven. If the comet's brightness curve was like that of Figure 6b, we could interpret the "three days" in Text 5 as referring to the briefer period comprising the few days near the maximum of the flare-up, when the comet would have reached magnitude -7 and may have been visible as it set in the NW at 14:34.

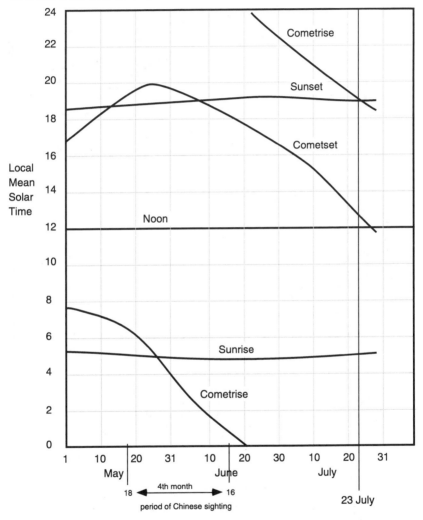

Figure 9: Rising and setting times for the Sun and Comet Caesar (the retrograde orbit with perihelion = 0.224 A.U.) from 1 May to 30 July 44 B.C., as seen from 34° N (the latitude of the Chinese capital rounded to the nearest degree). The time for local noon is also indicated.

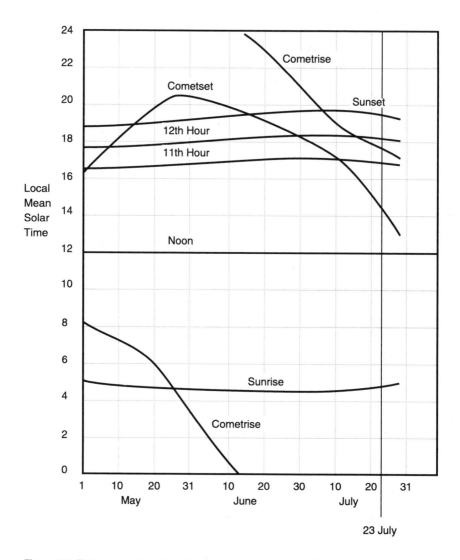

Figure 10: Rising and setting times for the Sun and for Comet Caesar (the retrograde orbit with perihelion = 0.224 A.U.) from 1 May to 30 July 44 B.C., as seen from 42° N (the latitude of Rome rounded to the nearest degree). The times for the beginning of the Roman 11th and 12th hours and local noon are also shown.

Identifying the Comet

Abstract

Given our estimated parameters for the comet of 44 B.C., we explain the difficulty of matching them with the parameters of known comets and so identifying Caesar's comet with one whose period is known.

It is interesting to consider the possibility that Caesar's comet has been observed at a later time. At first thought, one might take the period **P** of one of the comets in Marsden's catalogue that is known to have been at perihelion at time **T** and see if the arrival time after a number **n** of orbits (T - nP) is -43 years, which would put it near the Sun in 44 B.C. Unfortunately, cometary periods change, and after a few orbits the true arrival time could differ from this simple calculation by some years. Also, it seems likely that Caesar's comet is nearly parabolic (see p. 84), with a period over 200 years, and if it has been seen since, its period may very well not have been determined.

Another possibility would be to see if the parameters found above (Table 1, p. 126) match those for some comet listed in Marsden's catalogue. Unfortunately, the parameters of a comet are potentially subject to change over the course of time. Planetary gravitational attraction makes the longitude of the ascending node (Ω) and the argument of perihelion (ω) precess. If, however, a comet does not happen to make a close encounter with a planet, its perihelion distance and its inclination may remain relatively stable, and so it might be possible to match those two parameters of Comet Caesar with the corresponding parameters of a comet that has been catalogued. Given our value of roughly 109°±20° for the inclination of Comet Caesar, with 20° being the standard deviation, we might expect the true value to be within the range of 90° to 130°, and Marsden's catalogue lists 15 near-parabolic comets having inclinations within that range and a perihelion distance of less than 0.26 A.U. Starting from the modern observed positions of those 15 comets, it would be necessary to integrate numerically the orbits of each of them backwards in time to the year 44 B.C. and then compare their positions with our estimates for Caesar's comet. It would be an interesting exercise to do these integrations, but there can be very little hope of finding a match since the number of known near-parabolic comets is a very small drop in a very large bucket.

VII

THE INTERPRETATION OF CAESAR'S COMET
AS A POSITIVE OMEN

Statement of the Problem

We come now to the last, and perhaps most intriguing topic in our investigation of Caesar's comet: its interpretation by the Romans. What caused that comet to be treated so differently from the vast majority of comets and to be judged."beneficial for the world" (*salutare . . . terris*, Text 1A)?[1] Ordinarily, comets were viewed in antiquity (and until fairly recent times) as baleful signs, *omina dira*, portending the outbreak of war,[2] the demise of rulers,[3] or famine, plague and countless related sufferings.[4] Gundel in his Pauly article on comets was able to adduce no more than two or three other instances in which a comet received a positive interpretation in antiquity.[5] He also readily, and quite rightly, admitted that one has to

---

[1] It was unique in being honored in a Roman temple, the aedes Divi Iuli (Text 1A); an aureus and denarius issued by Octavian in 36 B.C. depict that temple with a star on its pediment: see Crawford no. 540 and plate LXIV. The temple stood facing the Capitoline, on the spot in the Roman Forum where Caesar's body had been cremated; it is alluded to in Text 30, line 842.

[2] *Div.* 1.18, *Nat. D.* 2.14; Tib. 2.5.71; Verg. *Georg.* 1.488; Manilius 1.866; [Sen.] *Octav.* 236.

[3] Lucan 1.529."a comet bringing change in regimes on Earth" (*terris mutantem regna cometen*); Tac. *Ann.* 14.22, 15.47; Suet. *Claud.* 46, *Nero* 36, *Vesp.* 23.4; Sil. Ital. 8.637."overthrower of kingdoms" (*regnorum eversor*).

[4] Isid. *Orig.* 3.71.16; Serv. on *Aen.* 10.272; specifically, storms (Sen. *Nat. qu.* 7.28.1; Claudian 33.234-36), drought and failure of crops (Manilius 1.894; Avienius, *Aratea* 1814-19; Serv. on *Aen.* 10.272), and earthquakes (Sen. *Nat. qu.* 7.28.1).

[5] Coll. 1149.65 ff: (1) the λαμπάς of 344 B.C. that accompanied Timoleon on his voyage to Sicily (Diod. Sic. 16.66.3; Plut. *Tim.* 8); (2) the comets that foretold the greatness of Mithridates at his birth and accession to the throne in the late second century B.C. (Justin 37.2.2-3); and (3) the comet that according to the Latin adaptation of Josephus (Hegesippus 5.44.1) was widely (*vulgo*) interpreted in A.D. 69 as a sign of

distinguish scholarly and astrological speculation concerning the positive import of comets under certain circumstances, from popular belief, which invariably looked upon comets as baleful. Allowance must also be made for flattery and the desire on the part of certain authors to interpret a particular comet in a way that would be pleasing to the Roman emperor.[6]

Our sources tell us, however, that in 44 it was not scholarly opinion but rather the common people (*vulgus*) who adopted the view that the comet was a sign of Caesar's apotheosis.[7] Significantly the common people are said to have arrived at this conclusion with encouragement from Octavian,[8] and we must also bear in mind that the extant accounts of Caesar's comet reflect primarily the view of the event that Octavian (Augustus) chose to present just over two decades later when he wrote his *Memoirs* (see p. 63). According to this Augustan tradition (Text 1), the people first interpreted the celestial phenomenon as a sign that Caesar's soul had been elevated to the ranks of the gods,[9] and then Octavian dedicated a statue of Caesar with a "star" (*sidus*)—not a "comet" (*sidus crinitum*)—above his head in keeping with the popular interpretation of the heavenly sign. This was *not*, however, the only interpretation of the comet that was current in 44 B.C., nor did Octavian necessarily welcome the view at first, as he did later in his *Memoirs*, that the bright light in the sky *was* a comet. (It was, after all, a star rather than a comet that was adopted by Octavian in 44 as a symbol for

---

coming freedom for the Jews, but not by the *prudentiores*, who regarded it merely as a sign of coming war. In Josephus (*BJ* 6.289) that comet is said to have been viewed as a harbinger of the destruction of Jerusalem. To these, add perhaps the "torch" (*fax*: on this term, see p. 93) of 204 B.C., one of the prodigies that encouraged the Roman people to expect victory in the Second Punic War (Livy 29.14.3). Recently, Hazzard 426-27 has speculated on very slender evidence that the comets of 210 and 204 B.C. may have been interpreted as positive omens by Ptolemy V Epiphanes. We thank Curtis Clay for directing our attention to Hazzard's article.

[6] So, for instance, Seneca asserted that the comet in A.D. 60 "blotted out the evil repute of comets" (*cometis detraxit infamiam, Nat. qu.* 7.17.2; yet it was followed by storms and earthquakes, 7.28.1-3). Calpurnius Siculus (Text 33) describes possibly this same comet as "shining with a serene and wholesome light" in contrast with the bloody and ominous comet in 44. Doubtless the same motive in part influenced the Stoic Chaeremon, who was one of Nero's tutors, to relate in a treatise on comets that some were a positive omen (Orig. *Contra Celsum* 1.59): see Schwyzer 62-63, who, however, speculates that Chaeremon may have found in one of his sources the concept that comets can occasionally be propitious signs.

[7] *Vulgus credidit*, Text 1; *persuasione vulgi*, Text 9.

[8] *Augusto persuadente*, Text 5; *persuasione Augusti Caesaris*, Text 8.

[9] Bömer 27-34, calls attention to the fact that in Augustus' *Memoirs* the comet was treated as a "sign" of Caesar's apotheosis, whereas in Ovid (Text 30), Baebius Macer (as reported in Text 6), Suetonius (Text 9), and Dio (Text 3) the star, or comet, is treated as identical with Caesar's soul (*anima*). That is, the earlier Roman interpretation soon yielded under Hellenistic influence to reinterpretation as a catasterism (καταστερισμός).

Caesar's new divinity, and according to Text 5, Octavian placed a star on his own helmet to advertise his connection with Caesar.) One attested rival tradition, which was clearly hostile to Octavian and his supporters, considered the heavenly body to be a comet that "portended the usual [baleful] things" (Text 3, cf. 6). Other persons offered interpretations that were positive, to be sure, but different from the one that Octavian ultimately adopted in his autobiography. For instance, there were those who speculated that the comet had been sent to add lustre to Octavian's renown (Text 6, on the authority of Baebius Macer), and even Octavian himself is credited with privately holding a somewhat similar view, although publicly he fostered the opinion that the comet signified the deification of Caesar (Text 1A).

Physical Appearance of the Comet and Contemporary Debate Concerning its Classification

In order to understand how the comet of 44 came to be treated so differently from most comets, it is helpful to take into account some of its unique features that have been revealed in our discussion of the sightings from China and Rome and the probable orbit. To begin with, we have seen that this comet apparently attracted little, or no notice among western observers when it approached perihelion in late May. Although we estimate that it was a bright object at that time, having a magnitude of perhaps -4 or -3, it may not have emerged far enough from the sunset skyglow to attract the attention of observers in Italy. Viewing conditions at that time could also have been a factor that hindered amateur skygazers (as opposed to trained professionals in China) from catching sight of Caesar's comet; the skyglow is likely to have been brighter and to have extended a bit farther than normal in the spring of 44 as a consequence of the volcanic veil produced by a recent eruption of Mt. Etna (see above, pp. 99-107). Later on, there is just a hint in two of our texts (Nos. 11, 12), as we have indicated (pp. 91-94), that the comet was possibly observed during the predawn hours in June or early July, but by that time it may have lost its tail and was presumably growing fainter in the starry sky. The evidence that it was seen at all by western observers before late July is extremely slim. Then suddenly, most likely because of a violent explosion of gases, it experienced a brief and very spectacular flare-up in late July. It became so bright, so suddenly, that it was visible in broad daylight. A comet of such a magnitude is extremely rare, as compared with ordinary, naked-eye comets of which as many as ten to twenty per century may be so bright that they are bound to attract the attention of even a casual observer.[10] Daylight comets, by contrast, occur on average not more than twice a century.[11]

In addition to this, the daylight comet of 44 just happened to coincide with games that were being celebrated not only in Caesar's honor but also in honor of the goddess Venus since, as we have argued in part one of this study, the games in 44 were most likely still being called the *ludi Veneris Genetricis*. Of course the symbol especially identified with the goddess Venus was a bright star, and the star of Venus had recently enjoyed considerable prominence on Rome's coinage under Caesar, serving to advertise the claim of the Julian gens that it was descended from the goddess Venus.[12] Schilling (322-23) draws attention to the role that the Augustan poets assign to Venus in accomplishing the astral apotheosis of Caesar (Texts 20, 24, 30). The role was a natural one for the putative ancestress of the Julian gens to play, and, we might add, it may well have been suggested almost immediately to observers of the comet in 44. As chance would have it, the planet Venus is likely to have been visible in the evening sky not long after the *sidus Iulium* rose at about the 11th hour in late July. On 23 July 44 B.C. the celestial longitude of Venus was 144.08° (*SKCL*), and so it was approximately 27° east of the Sun (117.28° Long.): see Figure 3, p. 88, for the positions of Venus and the Sun relative to each other. Therefore, the planet Venus should have been visible at twilight as the evening star if the sky was relatively free of the volcanic dust veil by that time of the year. Since Venus set on 23 July 44 B.C. at 20:51 (*SKCL*), an hour and 23 minutes after sunset (19:28), and had a magnitude of -3.3 (Houlden-Stephenson 140), we can be almost certain that the star of Venus would have been seen in the west for at least a short interval while the *sidus Iulium* was beaming in the NE. Surely

---

[10] Menzel-Pasachoff 384 estimate that a comet of the 1st or 2nd magnitude appears about once a decade. Actually the number may be slightly higher. If we take the final 270 years covered by Hasegawa's catalogue of naked-eye comets, the period for which we have the most complete records (A.D. 1701-1970), we find that 49 comets having a magnitude ≥ +2 were seen from the northern hemisphere, for an average of 18.1 such comets per century.

[11] Sagan-Druyan 124. Occasionally, however, this number may be exceeded; there were, for instance, two daylight comets back-to-back in 1882 (nos. I Wells and II) and in the 20th century there are those who may have witnessed as many as three daylight comets (1910 I, 1927 IX, and 1965 VIII); see Kronk under each of these years.

[12] See Weinstock 376-77 for a review of this coinage. Recently Molnar has speculated that the close identification of Caesar's fortune with the star of Venus may even lie behind the haruspex Spurinna's warning to Caesar that he should be on his guard during the 30 days ending with the Ides of March (Val. Max. 8.11.2). Molnar 8 points out that Venus experienced a heliacal setting on the Ides, a date on which Venus' powers were weakened by the position of Saturn, and Scorpio (a sign somehow significant for Caesar?) was setting (Pliny, *NH* 18.237).

this happy coincidence could not have been overlooked by those who wished to exploit the celestial phenomenon to argue that Caesar's soul was soaring to the heavens where it would join the ranks of the gods.[13]

The next important piece of evidence to take into consideration is that, according to several of our sources (Texts 3 and 6), an attempt was apparently made at the time of the observation of the *sidus Iulium* to deny that it *was* a comet. While some vehemently insisted that it was indeed a comet, portending the usual baleful events, others chose to interpret the bright burst of light in the sky as a new "star", not a comet. Its star-like appearance is, in fact, attested by Baebius Macer, who was most likely a contemporary observer; he describes the *sidus Iulium* as a "very large star surrounded with rays resembling streamers on a garland (*lemnisci*)."[14] It could very easily, therefore, have been mistaken for (or arbitrarily interpreted as) a star, instead of a comet. The dispute is not so surprising, given Seneca's remark that observers of comets frequently disagree concerning the shape of the object they are beholding because the sharpness of their eyesight varies (*Nat. qu.* 7.11.3). The recent observations of comet Hyakutake in March 1996 amply bear out Seneca's assertion: there were those who clearly saw Hyakutake's tail, while others could discern no tail at all but merely a "fuzzy ball".[15] Doubtless Comet Caesar did not have in July an extended "tail" of the type that is commonly associated with comets,[16] and yet the Greeks and Romans in their classification of comets did make allowance both for those that had tails coming to a point and for those that dispersed their rays more evenly on all sides.[17] Therefore, both schools of

---

[13] Significantly, Ovid (*Met.* 15. 519-20) credits Venus *and* Caesar's son Octavian with bringing about Caesar's deification, and according to a theory that attributed comets to the influence of the planets, "comets produced by the planets Venus or Jupiter foretell propitious happenings" (*nam si de Venere aut Iove fiant, optima praenuntiant*, Servius on *Aen.* 10.272).

[14] Text 6; cf. Text 13 "surrounded on all sides by hair". For illustrations of *lemnisci*, the ribbons or streamers that were sometimes attached to garlands, see Dar.-Sag. I.2 1523 *corona* fig. 1978 and III.2 1100 *lemniscus* fig. 4436.

[15] Personal communication with Brian Marsden.

[16] Sometimes, however, in modern literature the comet that appeared during Octavian's games is erroneously said to have had "a long tail": e.g., "ein mächtiger Komet mit langem Schwief" (Gardthausen I.1 54 and Ehrenwirth 61, employing the identical words); "von acht Ellen Länge" (K. Fitzler-O. Seeck, *RE* 10.1 [1917], 218, appealing to the Chinese report without taking into account that it concerned a much earlier sighting in May-June).

[17] According to Aristotle (*Mete.* 1.7.344a 23-24), if the flame extended equally in all directions it was called "long-haired" (κομήτης), and if it extended lengthwise, it was called "bearded" (πωγωνίας). Later the term *pogonias* apparently came to be restricted to those comets whose flame hung down like a beard (Pliny, *NH* 2.89), as distinguished from

thought could present their rival interpretations of the celestial phenomenon in 44 with ample justification; it was arguably a star *or* a comet.

One of the spokesmen for the comet-interpretation, we are informed, was the haruspex Vulcanius, who regarded the comet as a sign that the "ninth age" was ending and the "tenth" beginning (Text 6).[18] Comets and meteors, as well as lightning, appear to have figured in the Etruscan art of haruspicy and seem to have been treated in the so-called *Tarquitiani libri*, in which the principles of the Etruscan art were set out in a Latin translation.[19] To judge from at least one later reference, the haruspices did not regard these signs as propitious.[20] Presumably Vulcanius referred in his prophecy to the Etruscan belief in the ten ages that were to comprise the span of that nation's existence.[21] If so, then the new tenth age that was supposedly ushered in by the appearance of the comet in 44 could be expected to complete a cycle of steady decline, leading to oblivion at its end.[22] In other words, Vulcanius' prophecy was an extremely gloomy pronouncement

---

those whose flame came to a point and those that "scatter their rays on all sides like hair" (*qui* [sc. *flammam*] *undique circa se velut comam spargunt*, Sen. *Nat. qu.* 7.11.2; cf. 7.6.1 "pour their gleam on all sides", *ardorem undique effundunt*).

[18] According to Etruscan doctrine, when one "age" (*saeculum*) reached its conclusion and another commenced, the gods sent signs because the length of these ages varied and it was difficult for man to know where one age left off and another began. The length of each *saeculum* is said to have equaled the life span of the oldest member who had been alive (or was born) when the age began (Censorinus 17.5).

[19] See Bouché-Leclercq 550 and his n. 1. On the *libri Tarquitiani*, see Thulin, *RE* 7 (1912) 2464.

[20] In A.D. 363 the haruspices advised the emperor Julian in the course of his Persian expedition to postpone all undertakings following the appearance of a meteor (*fax cadenti similis*), but the emperor disregarded this advice and was killed soon afterwards in a skirmish (Amm. Marc. 25.2.7). Among the fearful portents that appeared to Jovian at Antioch, it is said that "comets were seen during the daytime" (*visa sunt interdiu sidera cometarum*, Amm. Marc. 25.10.1-2).

[21] Censorinus 17.6. Nilsson, *RE* 1A (1920) 1701-1708, rightly insists that the tenth age of Vulcanius' prophecy should be viewed in the context of Etruscan, not Hellenic or Roman theory.

[22] There is some problem in making Vulcanius' "tenth age" fit with what we know about the length of each of the preceding ages, ranging from 100 to 123 years each according to Censorinus (17.6). Since the commencement of the ninth Etruscan age seems to have been marked by portents in 88 B.C. (Plut. *Sull.* 7), that age could scarcely have run out a mere forty-four years later, at the time of the comet in 44. Weiss 216 n. 47 speculates that perhaps Vulcanius departed from the Etruscan scheme and adopted Varro's date of 149 as the beginning of the previous *saeculum*; but see preceding note. Be that as it may, to judge from the reference to the prevalence of greed (*avaritia*) in the recent, eighth age in the so-called "Prophecy of Vegoia" (ap. Lachmann 350-51), each age was apparently expected to be more decadent than its predecessor. See Harris 31-40, for bibliography and a discussion of the authenticity of "Prophecy of Vegoia", esp. 36-37 for the problems posed by Vulcanius' prophecy.

regarding the end of the world, quite in keeping with the usually baleful nature of comets.[23]

We can confirm that this is undoubtedly the correct interpretation of Vulcanius' prophecy by comparing it with an apparent doublet that we find in Appian. In both instances, the prophet is an Etruscan haruspex (anonymous in Appian), and the prophecy is uttered in response to ominous occurrences. In Appian, the context is the dire portents that attended the formation of the so-called Second Triumvirate in 43 B.C.[24] Several of those prodigies (dogs howling, monuments struck by lightning, and especially the fearsome appearance of the Sun) recall portents that are also said to have occurred in 44, the year of Vulcanius' prophecy (Dio 45.17.2-7; Obsequens 68). The prophecy of Appian's aged haruspex foretold the return of the age of kings and the enslavement of the Roman people; its content recalls the oracles that, according to Dio (45.17.6), circulated in 44, announcing the end of democracy (δημοκρατία) at Rome. In both instances, the prophets sealed the veracity of their pronouncements by dying on the spot, a common motif in apocalyptic prophecies.[25] We should note, however, that the death of Vulcanius and of the anonymous haruspex in Appian cannot be wholly explained as simply a traditional feature of such prophecies. The latter chose to take his own life, we are informed, so as to avoid witnessing the enslavement of the Roman people and being himself a slave, while Vulcanius is said to have collapsed and died in the course of uttering his vaticination "because he was announcing, against the will of the gods, matters that lay hidden."[26] The way

---

[23] The notion that a comet may be an omen signifying the impending destruction of the world is found once or twice elsewhere: (1) perhaps in Manilius 1.903 if we take *finemque minata est* as "threatened her own (i.e., nature's, the world's) destruction" (Goold translates "our [i.e., the Romans'] destruction"); and (2) in a simile in Sil. Ital. 1.464, "threatens an end to the world" (*terrisque extrema minatur*).

[24] Appian 4.4.15. Hahn 243-44, perceptively calls attention to this doublet for the light it sheds on the intent of Vulcanius' pronouncement. Weinstock 195 briefly touches on the prophecy of Appian's haruspex in 43 but sees the death of both haruspices as intended chiefly to mark the close of a *saeculum*, without relating this feature of both prophecies to the gloomy, versus hopeful, outlook for the future.

[25] This motif is found, for instance, in the so-called *Oracle of the Potter* and in the *Oracle of the Lamb*, which belong to a long tradition of such prophetic texts in Egypt, and which in their Greek form stretch back to at least the fourth century B.C. See Koenen, *ZPE* 2, for text and interpretation, and *ZPE* 3 and *ZPE* 13 for supplements. We thank Ludwig Koenen for discussing with us the *Oracle of the Potter* and related texts. He is not, of course, to be held responsible for any of the views expressed here.

[26] *Quod invitis diis secreta rerum pronuntiaret*, Text 6: since the reason why Vulcanius was doomed to death is incorporated into the indirect statement by means of the subjunctive *pronuntiaret* (sub-oblique), the reason is represented as the one given by Vulcanius himself.

in which Vulcanius' death is described by the victim himself as following from his revelation of what the gods wished to conceal provides further confirmation that Vulcanius must have predicted hardships. As Hahn (242) has pointed out, there would scarcely have been any motive for the gods to punish Vulcanius if his intended message had concerned the coming of a new Golden Age, as modern scholars sometimes assume incorrectly (e.g., Weinstock 195, 371).

It is likely, however, that already in 44, or soon thereafter, the intent of Vulcanius' message became a subject of dispute, leading ultimately to a radical reinterpretation of what the new, tenth age would bring. What appears to be one such reinterpretation of the nature of Vulcanius' tenth age is preserved in a nearly contemporary text. In the *Fourth Eclogue*, which foretells the commencement of a new and glorious age in the consulship of Asinius Pollio (40 B.C.), Virgil informs us of a prophecy attributed to the Sibyl of Cumae, according to which the tenth age would usher in a renewal of the world under the beneficent rule of the Sun.[27] This doctrine (Greek, rather than Etruscan) clearly played no part in Vulcanius' interpretation of the comet as a sign that the tenth age was about to begin (Norden 15 n. 1), but it may well have appealed to Octavian and his supporters, a circle that was to include Virgil. The Sibylline version of the tenth age provided a ready means of modifying Vulcanius' gloomy and negative prophecy, offering in its place a hopeful message that the tenth age was to be a new Golden Age (*aurea aetas*) of peace and renewal. The plausibility of this rival interpretation could only grow stronger with the passing years in response to Octavian's success in putting an end to the cycle of Rome's bloody civil wars.

It should come as no surprise that the popular fancy was receptive to the notion that a new age was on the verge of dawning in 44 B.C. The timing for such a prophecy was ideal. The last set of Secular Games, celebrating the dawning of a new age, or *saeculum*, had been held in 146 B.C. (postponed, it seems, from 149).[28] There are signs that Caesar may have been planning a

---

[27] "Now the last of the ages in the Cumaean prophecy has arrived" (*ultima Cumaei venit iam carminis aetas*, Verg. *Ecl.* 4.4, on which Servius writes: "[the Sibyl of Cumae] divided the ages according to designations by metals, stated who ruled in each age, and desired the Sun to be the last, that is, the tenth ruler" (*saecula per metalla divisit, dixit etiam quis quo saeculo imperaret, et Solem ultimum, id est decimum, voluit*). It would not be surprising if the appearance of the comet in 44 led to a consultation of the Sibylline Books (*libri Sibyllini*) by order of the Senate (Coleman 130), but behind Virgil's conception of the tenth, Golden Age there clearly seems to be present an element of "eastern", as opposed to "western", ideology, possibly based on a source reflecting Jewish eschatology: see Nisbet 60-61.

[28] Weinstock 193; according to Varro, those games had been instituted in 249 B.C. following a consultation of the *libri Sibyllini* and were to be held every one hundred years

celebration to mark the advent of a new age, perhaps to follow his anticipated victory over the Parthians (Weinstock 191-7), but those plans were naturally cut short by his murder on the Ides of March. The chance occurrence of the comet, however, and Vulcanius' prophecy only served to sharpen the expectation that a new age was being ushered in. Coinage in the following years adopted symbols intended perhaps to suggest the advent of the "great year" (*magnus annus*) when the heavenly bodies would return to the positions they occupied at the creation of the world and so mark the beginning of a new cycle.[29] In the end, Augustus held the long-anticipated set of Secular Games in 17 B.C. after he had settled the Parthian question, and at that time we find on the coinage of the moneyer Marcus Sanquinius a comet above the head of a youthful figure who has sometimes been identified as the deified Julius Caesar or his Trojan ancestor Iulus (Ascanius) son of Aeneas, but may well be the personification, or *Genius*, of this new Augustan *saeculum* (Boyce 6-7).

Even any lingering hint in Vulcanius' prophecy of suffering and turmoil could be turned to advantage and made to fit the Augustan reinterpretation; gloomy elements form a common motif in prophecies concerning the advent of a new and glorious age under the beneficent rule of a divinely appointed Savior. Typically such prophecies assert that before the glorious renewal of the world can take place, a period of suffering and trial must first be endured.[30] For Octavian's contemporaries, such a period of suffering could easily be found in the protracted series of civil wars that finally ended in 31 B.C. with Octavian's victory at Actium over Mark Antony and the Egyptian queen Cleopatra.

This reinterpretation is almost certainly the way in which Octavian must have presented Vulcanius' prophecy in book two of his *Memoirs*, where it found a place according to Servius Auctus (Text 6). By contrast, the account of Vulcanius' prophecy that is preserved in Text 6 appears to have been

---

(Censorinus 17.8, 11; cf. Pseudo-Acron on Horace, *Carm. Saec.* 5).

[29] Cic. *Nat. D*. 2.51; cf. Censorinus 18.4. The moneyer Publius Clodius (42 B.C.?) issued both aureii and denarii having the radiate head of Sol on their obverse and a crescent Moon and five stars (representing the planets) on their reverse: Crawford no. 494/20a & b, 21, p. 511 for discussion and bibliography.

[30] This motif is found both in the *Oracle of the Potter* and *Oracle of the Lamb*; on the former, see Koenen, *ZPE* 2.180. On the 900 years of misfortune that were to precede the dawning of a new and glorious era of salvation as related in the *Oracle of the Lamb*, see Koenen (1984) 10-11. We thank Prof. Koenen for sharing with us that portion of his unpublished Habilitationsschrift (Appendix 3) in which he discusses an oracular tradition in Rome of nine *saecula* that is reminiscent of the 900 year cycle in the Egyptian prophecies.

drawn not from Octavian but from Baebius Macer.[31] This circumstance will account for the antithesis that is drawn between the view (of Octavian and his supporters) that the heavenly body was a star and Vulcanius' assertion that it was a comet ("*But* Vulcanius stated that it was a comet," *Sed Vulcanius . . . dixit cometem esse*, Text 6). As we can see from Octavian's act of placing a golden star above the head of Caesar's statue in 44 (Texts 1, 3, 6, cf. 5), Octavian must have been among those who at first disagreed with Vulcanius and chose to interpret the bright light in the NE as a "star." Both on the coins of the following decades and in Augustan literature down to Ovid (Texts 20-24), whenever allusion is made to the comet of 44 in the context of Caesar's apotheosis, it is always portrayed as a "star", not a comet.[32] By contrast, whenever the theme of Caesar's apotheosis is absent, the same phenomenon is described as a comet with the usual baleful implications, and it is included in the list of dire portents that followed the Ides of March (Texts 31-33). In other words, in the latter texts the comet is treated as a sign of the impending civil wars that were destined not to come to an end until Octavian's victory at Actium ushered in the *pax Augusta*. Interestingly, we find both interpretations of the comet of 44 in Virgil, who classifies it as a "star" (*astrum*, *sidus*) when celebrating the deification of Caesar (Texts 20, 21), or a "comet" (*cometes*) when mourning Caesar's death (Text 31).

Apparently Octavian himself eventually modified his view of the comet, abandoning his initial stance that the heavenly body was a star, not a comet. At least by the time when he composed his *Memoirs* (circa 24 B.C.), he was prepared in book two of that work to describe the *sidus Iulium* as a *sidus crinitum*, i.e., a "hairy star" or "comet" (Text 1).[33] Augustus' acceptance of

---

[31] This is a significant detail that has frequently been overlooked. Scott 259 n. 9, for instance, incorrectly states that "the account of Vulcanius, at least, was taken [by Servius] from the second book of Augustus' *Commentarii*."

[32] For a review of the coinage, see Weinstock 377-81 and Plate 28. See Scott for a discussion of both the coinage and literary allusions to the *sidus Iulium*. Scott, however, fails to call attention to the pronounced shift in the imagery, from star to comet, that is to be seen in the comet coins of 17 B.C. and in Ovid's portrayal of the apotheosis.

[33] Ironically, the most recent discussion of this passage (the sole surviving fragment of the *Memoirs* in Augustus' own words) neglects to take into account this very significant change in terminology: Bramble 492 states that the passage gives an account of the "*star* [emphasis added] which heralded Caesar's apotheosis." The term Octavian chose, "*sidus crinitum*," is without precedent as a term for a comet. It is also a striking expression because on its own, *sidus* in the singular ordinarily denotes a constellation rather than a single star (*stella*): OLD s.v. *sidus* 3a; cf. Servius on *Aen.* 8.681: "to show him honor the poet calls a single star a *sidus* although a *sidus* consists of many stars" (*Honorifice autem poeta unam stellam 'sidus' dixit, cum sidus ex multis stellis constet.*). While *sidus* is used freely by authors writing after the publication of Augustus' *Memoirs* (Texts 21, 22, 25-27) to describe the comet of 44 as a "star" (becoming a stock means of alluding to

this view will account for Ovid's portrayal of Caesar's apotheosis as the birth of a comet in his *Metamorphoses* (Texts 29, 30), written between A.D. 2-8, long after Augustus' *Memoirs* and the comet coins of 17 B.C.,[34] whereas Virgil, Horace, and Propertius writing in the earlier period adopted the star imagery exclusively (Texts 20-24, composed between circa 42 and the late 20's B.C.). Undoubtedly when Augustus wrote his *Memoirs*, Vulcanius' prophecy was given the more positive interpretation described above and made to concern the return of a Golden Age.

Astrological Interpretations of Comets[35]

If we ask how Augustus was able to adopt the view in his *Memoirs* that the *sidus Iulium* was indeed a comet (as Vulcanius had claimed) but one that was a propitious omen (not baleful, as comets usually were), we shall find the most promising answer to this question in texts that discuss astrological interpretations of comets.[36] We are informed by Pliny (*NH* 2.92-93), for instance, that people regarded comets as sending different messages according to their shape, the quarter of the sky in which they were observed, the direction in which they pointed, and their linkage with other heavenly bodies (cf. p. 139.13). As chance would have it, we possess quite a full astrological interpretation of a particular type of comet that is specifically said to be the type that appeared at the beginning of Octavian's political career (Text 13)—i.e., presumably in July 44, when his first public act was to hold the games in honor of Venus and Caesar. This text and the balance of our

---

deifications in the imperial family, *OLD* s.v. *sidus* 3c), the Latin sources (other than Augustus) that call the heavenly body a "comet" either use the Greek term *cometes* (Texts 1A, 2, 6, 13, 31-33) or refer to it as a *stella crinita* (Texts 4, 9) or *stella comans* (Text 29).

[34] For the change marked by the coinage in 17 on which the image of a comet replaced that of a star, see Weinstock 379. The only other representation of the *sidus Iulium* as a comet on an object that *may* belong to the Augustan period is found on an incised carnelian, which was in a private collection before World War II: published only once, by Pesce 402 and Tav. 1.1. It shows Caesar's head in profile, wreathed with laurel and beside it a comet at the level of his brow. This gem is not taken into account by Weinstock (presumably because he could not examine it so as to judge its authenticity, although he cites Pesce's article), and our attempt to locate the present whereabouts of this stone by writing to numerous scholars in Europe and America has proven unsuccessful. Gems more typically depict the *sidus Iulium* as a star rather than a comet: e.g., Bernoulli 151-52, Vollenweider I 196.

[35] We thank our colleagues Matthew Dickie, Lambros Missitzis, and John Vaio for discussing with us the interpretation of a number of passages from John Lydus and Hephaestion Thebanus that bear upon this topic. They are not, of course, to be held responsible for any of the conclusions drawn here.

[36] For a survey of this literature, see Bouché-Leclercq 357-61 and Gundel 1153-64.

information come from late sources, but most likely the gist of this lore goes back to an astrologers' handbook that appears to have been put together in the second century B.C. and circulated under the names Nechepso and Petosiris.[37] The descriptions in these astrological texts tally quite closely with what we have determined from our other sources were the probable shape and location of Comet Caesar. We are told, for instance, that if a comet is surrounded with rays shooting out in all directions and "looks upon the west, the land of Italy will achieve its fondest wishes" (Text 13). This interpretation of what appears to be the comet of 44 B.C. is attributed to Avienius (fourth century A.D.) and may be compared with what John Lydus (flourished circa 560 A.D.) tells us about a type of comet called κομήτης, which is said to bring good fortune to the region on which it beams with its bright and silvery rays.[38] Both Lydus and Hephaestio of Thebes (writing circa 415 A.D.) associate that type of comet with the planet Jupiter, and they stress the brilliance with which it shines, bringing good fortune on the region it beholds. Hephaestio adds that it appears to have in it the face of a man or of a god.[39] Both Greek sources also state that that type of comet is especially

---

[37] See Pieper, *RE* 16 (1935), 2160-67, esp. 2163-64 for the probable date of composition. Recently, however, David Pingree 548 has questioned whether the astrological texts concerning comets in Lydus, Hephaestio, and Avienius necessarily go back to a source as early the astrological handbook attributed to Petosiris. He points out that unlike the other material in the fragments assigned to Petosiris, there has not yet been discovered in the cuneiform tablets a Babylonian source for the material concerning comets.

[38] "If it is turned the other way [from Persia in the E], having the E behind it, and looks to the W, the Romans shall have boundless prosperity. . . . And the Roman nation shall have an abundance of all good things at once. There will be games on a magnificent scale and lavish gifts. Splendid circensian games will be held to the gratification of the common people. In short, there will be universal joy in all doings." (εἰ δὲ ἀποστρεφόμενος καὶ κατὰ νῶτα ἔχων τὴν ἀνατολὴν ἐπὶ τὴν δύσιν ὁρᾷ, ἄπειρος ἔσται τοῖς Ῥωμαίοις εὐδαιμονία· . . . πάντων τε ὁμοῦ τῶν ἀγαθῶν ἀφθονία τῇ Ῥωμαίων πολιτείᾳ ἔσται· θέαι τε ἄφθονοι καὶ δωρεαὶ φιλότιμοι καὶ ἱππικοὶ ἀγῶνες σπουδαιότεροι μετὰ ψυχαγωγίας τοῦ πλήθους ἐπιτελεσθήσονται καὶ ἁπλῶς εἰπεῖν τοῖς πράγμασι καθολικὴ ἔσται ἱλαρία, *De ostentis* 15). The reference to games—circensian and otherwise—and to donatives is reminiscent of conditions in July 44 when Octavian held his games and began paying the legacies provided for the Roman people by Caesar's will (see above, p. 46). The only other known historical context in Roman times when games may possibly have been linked with the appearance of a comet is the celebration of the *ludi Saeculares* in 17 B.C., and it is likely that that comet (if it indeed existed: see above, p. 65.12) was viewed as a return of the *sidus Iulium*.

[39] "And it has on it a face with human features, as it seems of a god" (ἔχει δὲ ἐν ἑαυτῷ ἀνδρεῖον πρόσωπον, ὡς δοκεῖν θεοῦ, *Apotelesmatica* 1.24); cf. Pliny, *NH* 2.90, "and there occurs a radiant comet with silvery hair, so shiny that it can scarcely be looked at. It displays on it the likeness of a human face" (*fit et candidus cometes argenteo crine*

propitious when Jupiter, the planet with which it is associated, is in Cancer, or Scorpio, or Pisces, but we can determine that this condition was not fulfilled in 44.[40] Still, it is texts such as these that may well have been used in 44 B.C., or later, by Octavian to turn this potentially very negative event into a sign that was favorable to himself and the Roman people.

The SIDUS IULIUM and Octavian's Birth Sign Capricorn

It only remains now to consider the cryptic remark that Pliny (Text 1 A) attributes to Octavian and claims was Octavian's private view of the comet's significance: "that the comet had come into being for him and that he was coming into being in it" (sibi illum [sc. cometen] natum seque in eo nasci). The difference in tenses between natum and nasci makes it clear that the one event (the metaphorical birth of Augustus) followed the other (the "birth" of the comet). Modern scholars have generally professed puzzlement or interpreted this enigmatic remark as a comment on Octavian's confidence in his future. "Whatever these words mean" was H. J. Rose's despairing comment (p. 191), while according to Wagenvoort (18), the words undoubtedly had reference to a "mystic Theologumenon" and were to be explained by the assumption that "Augustus was expecting a παλιγγενεσία (rebirth) which signified the incarnation of some divinity in him."[41]

In our view, there is another, more promising explanation of this remark that emerges from a closer examination of the context in which it was made. As we have indicated, both astrologers and haruspices are likely to have been consulted at the time of the appearance of Caesar's comet, and a variety of conflicting opinions were doubtless competing for acceptance. On top of this, Suetonius (Aug. 94.12) informs us that the astrologer Theogenes had

---

ita refulgens ut vix contueri liceat, specieque humanae faciei effigiem in se ostendens).

[40] Pisces is one of the Houses of Jupiter, and the Exaltation of Jupiter is in Cancer (15°), these two constellations forming with Scorpio a trigon. Since Jupiter was at 300° R.A. in late July 44, it was on the cusp of Aquarius, or well within Aquarius, if we begin this sign 8° earlier according to the Mesopotamian System B, which was the one generally preferred by ancient astrologers (antiqui astrologi) according to Columella (Rust. 9.14.12). Nor was Jupiter in one of those three constellations in 17 B.C. either during the Secular Games (31 May - 3 June, with supplementary ludi 5-12 June), or in the months leading up to those games, which were perhaps held in response to the appearance of a presumed return of the sidus Iulium (see above, p. 65.12): Jupiter moved from Aries in Jan. to Taurus by June.

[41] Alföldi (1930) 380-81 adopted a similar view, calling attention to the belief later in antiquity that each new emperor ushered in a new and better age (saeculum) as a savior of his nation, perhaps personifying a renascent Jupiter (Iuppiter crescens). Weinstock 371 states more generally that Augustus signified by these words that the comet "portended his [Augustus'] own rise and his era."

recently cast Octavian's horoscope before he left Apollonia (in Illyria) in late
March 44, and Octavian presently (*mox*) came to have such faith in his
destiny that he published his horoscope and issued silver coins bearing the
sign Capricorn "under which he was born" (*quo natus est*).[42] Such coins do
indeed exist, and both Manilius (2.507-509) and Germanicus (*Aratea*
558-60) treat Capricorn as Augustus' natal sign, but modern scholars have
questioned how this could have been so. Since Suetonius states elsewhere
(*Aug*. 5) that Octavian was born shortly before sunrise (*paulo ante solis
exortum*) on *IX Kal. Oct.* (= 23 Sept. in the Julian, or 22 Sept. in the pre-Julian
calendar), and since in 63 B.C., the year of Augustus' birth, the civil calendar
appears to have been in close agreement with the Julian calendar,[43] the Sun
would have been in Libra on the day of Augustus' birth.[44] Therefore, Libra, not
Capricorn, must have been the constellation that was horoscoping (i.e.,
sitting on the eastern horizon) at Octavian's nativity, and some texts do
seem to treat Libra as Augustus' sign.[45] Various explanations have been
proposed to account for the importance that Capricorn obviously enjoyed in
Augustus' view of his horoscope. Possibly it was his Moon sign,[46] the sign of
his Lot of Fortune,[47] or, as recently re-asserted by Pierre Brind'Amour
(62-76) and Glen Bowersock (385-87), was the sign of his conception.[48] For
the purposes of our argument, it makes no difference which of these
explanations is accepted, and it is well to bear in mind that ancient
astrologers may have applied a variety of formulas to present their clients
with a horoscope that would be pleasing to them.[49] The indisputable fact

---

[42] Cramer 83 assigns the publication of Augustus' horoscope to A.D. 11, but the
evidence of the coins depicting Capricorn points to a date in the 20's B.C. The earliest of
these coins appears to have been issued in 28 B.C. (see Kraft 23), just a few years before
the publication of Augustus' *Memoirs*.

[43] See Brind'Amour 56-62 on the state of the calendar in 63 B.C.; he concludes (p.
71) that 22 Sept. of the Roman civil calendar = 22 Sept. of the Julian and that the Sun
occupied Libra from 17 Sept. to 17 Oct. Previous calculations for the year 63 have yielded
the following results (*SKCL*): 22 Sept. civil = 21 Sept. Julian (Holzapfel-Groebe); = 20
Sept. Julian (Soltau and Unger); = 29 Aug. Julian (LeVerrier).

[44] See Barton 41 fig. 1 for Augustus' natal horoscope based upon the calculations of
Brind'Amour (72).

[45] Verg. *Georg.* 1.32-35; Manilius 4.546-51; *De duobus signis* in *Anthol. Lat.* 43, ed.
Baehrens, *Poet. Lat. Min.* IV.144. The interpretation of Manilius 4.773-77 is disputed,
some seeing in that passage an allusion to Augustus (Brind'Amour 67-69), others to
Tiberius (Housman 112-13).

[46] So Housman: this view has most recently been adopted by Getty 101-107 and
Goold 284 note b.

[47] So Gundel (1926) 313-20.

[48] This same view was advanced by Bouché-Leclercq 373-74 at the turn of the
century.

[49] Barton 39-40 has pointed this out, arguing that there may not be a single reason

remains, however, that Capricorn was regarded by Octavian himself, for whatever reason, as the "sign under which he was born" (Suet. *Aug.* 94.12).

If we turn our attention to Figure 11, we shall see at a glance a significant feature of the late afternoon sky on 23 July 44 that has hitherto not been taken into account by scholars.

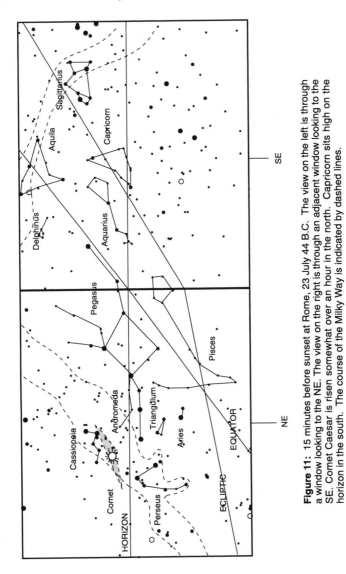

Figure 11: 15 minutes before sunset at Rome, 23 July 44 B.C. The view on the left is through a window looking to the NE. The view on the right is through an adjacent window looking to the SE. Comet Caesar is risen somewhat over an hour in the north. Capricorn sits high on the SE. horizon in the south. The course of the Milky Way is indicated by dashed lines.

behind the selection of Capricorn as Augustus' natal sign. See Abry 113-21 for an attempt to explain Augustus' choice to emphasize Capricorn over Libra.

During the period immediately preceding sunset, as shown in Figure 11, Capricorn was all but fully risen above the eastern horizon and for well over an hour before this time had been occupying the region where the ecliptic meets the horizon, a position known as the Ascendant which imparted to that constellation special significance in astrological terms. (As they rise, the degrees of the ecliptic "assume their special powers": *ortu / accipiunt proprias vires*, Manilius 4.503-504.) If we ask ourselves what the position of the stars would have been in the eleventh hour during which the comet is said to have risen, we shall find that the constellation Capricorn was indeed horoscoping (i.e., rising above the eastern horizon) during approximately the latter two thirds of that hour. This fact is revealed by the following table (Table 2) which gives the rising and setting times for nine of the stars belonging to the constellation. As we can see, α *Capricorni* rose at XI:20, while ζ *Capricorni* rose at the end of the twelfth hour (XII:58).

Figure 12. Capricorn

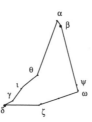

	Rise (LMT)	Set (LMT)	Rise (Roman)
Sun	4:38	19:27	I :00
α	17:24	3:30	XI :20
β	17:35	3:22	XI :29
θ	18:37	3:46	XII :19
ψ	18:45	2:45	XII :26
ι	18:56	4:02	XII :35
ω	19:00	2:39	XII :38
γ	19:16	4:18	XII :51
δ	19:22	4:27	XII :54
ζ	19:24	3:35	XII :58

Table 2: Rising times of nine stars belonging to Capricorn, on 23 July 44 B.C. The calculations were done using *SKCL*. One Roman hour equals 74.083 standard minutes on this date. The notation XII:26 signifies 26 Roman minutes after the beginning of hour XII. One Roman minute equals 1.235 standard minutes on this date.

We can now clearly see for perhaps the first time how Octavian may have concluded that the comet, or bright star, in the NE was to be interpreted as a sign of his birth. Given the important role played by Capricorn in Augustus' horoscope, if that constellation was the ascendant sign when the comet rose, Octavian could easily (with, or without, the encouragement of an astrologer) have read into that daylight prodigy a message intended for him personally, even connecting it with his recent adoption into the Julian family as Caesar's son. It must have seemed providential that the *sidus Iulium* appeared so soon after Octavian took his new name "C. Iulius C. f. Caesar."[50] Of course, this was Octavian's private view of the comet, not the one that he fostered publicly (Text 1A). Therefore, we should not expect Capricorn and a symbol of the comet to be prominently displayed in the official iconography, although there is one issue of denarii on which Capricorn appears on the reverse, bearing the legend "Augustus" below and having a star above, to the right, which may be intended as an allusion to the *sidus Iulium*.[51]

Finally, there is one further way in which the ascendant position of Capricorn on 23 July may have contributed to the interpretation of the comet of 44. We know that Capricorn later came to be regarded as the "Gate of Souls" by which the souls of the dead returned to "the seat of their inherent immortality" (*in propriae immortalitatis sedem*) and "to the number of the gods" (*in deorum numerum revertuntur*).[52] This doctrine was held by the Neoplatonists and was current at least by the latter half of the second century A.D. (and probably a good deal earlier).[53] What we can say for

---

[50] On the nomenclature of Octavian in the letters and speeches of Cicero in 44 and 43, see Shackleton Bailey, *Two Studies* 75. See Schmitthenner 58-59 for how the adoption and change of name is reported in our other sources. Octavian's friends were already addressing him as Caesar in April 44 (*Att.* 14.12.2), a month before Octavian appeared in Rome in the presence of Gaius Antonius, the acting praetor urbanus, to accept formally his inheritance from Caesar. His new name followed as a consequence of accepting this inheritance since it was written into the terms of Caesar's will: see Schmitthenner, 34-35, 49-51.

[51] *RIC*[2] 542: on the obverse is a youthful head of Augustus laureate, facing right; the mint is uncertain. It is classified as a rare issue, "R3" (= "up to about 10 specimens" known to exist).

[52] Macrobius (late 4th/early 5th cent. A.D.), *Commentarii in Somnium Scipionis* 1.12.2.

[53] Numenius fr. 34 (Budé ed. 1973, Édouard Des Places). If the frequent depiction of Capricorn on gravestones may be connected with this notion concerning the role of Capricorn as the "Gate of Souls", then the concept can be traced back to the 1st century A.D. (Barton 50). Possibly this idea can be detected earlier still, at the very beginning of the 1st century A.D., in verses 558-60 of Germanicus' *Aratea* which credit Capricorn with conveying the soul of Augustus to the stars: "Before the eyes of dumbstruck nations and

certain is that the Romans of Cicero's generation were familiar with the concept that the souls of great men might return to the heavens, whence they had come.[54] Furthermore, according to one version of this doctrine, the Milky Way was the abode to which the souls of heroes returned,[55] and that is precisely where the *sidus Iulium* was situated, if we are correct in concluding that it rose in the NE, in the vicinity of the constellation Cassiopeia, through which the Milky Way passes (see Figure 10).[56]

The role assigned to Capricorn by the Neoplatonists would seem to complement nicely the belief that the souls of heroes could ascend to heaven; no doubt Capricorn came to be viewed as the "Gate" for this ascent thanks to the association of that constellation with the winter solstice and the "rebirth" of the Sun in that season of the year, one of the positive features of the sign that may have encouraged Augustus to adopt Capricorn as his personal emblem. Capricorn is already firmly connected with those ideas in the Augustan poet Manilius (Abry 118-19), and one cannot help but wonder whether some of this thinking may lie behind the ready acceptance of the belief that the *sidus Iulium* signified the reception of Caesar's soul "among the spirits of the immortal gods" (*inter deorum immortalium numina receptam*, Text 1; cf. *in deorum numerum relatus est*, Text 9 and Macrobius quoted above). If so, this positive contribution of Capricorn to the message that Octavian wished to spread among the masses may have offered one

your own quaking fatherland, Augustus, this sign [Capricorn], which gave you birth, carried your divine majesty on its body into the heavens and restored it to the engendering stars" (*hic, Auguste, tuum genitali corpore numen / attonitas inter gentis patriamque paventem / in caelum tulit et maternis reddidit astris*). At the time of Augustus' death at the 9th hour on 19 Aug. A.D. 14 (Suet. *Aug*. 100.1), none of the five planets or Sun or Moon was in Capricorn, nor did Capricorn occupy one of the four Cardinal positions. Therefore, other than being the emperor's birth sign (styled *genitali*), there appears to be no particular reason for the poet to assign to Capricorn the role of restoring late emperor's soul to the heavens unless Capricorn was already coming to be regarded as the "Gate of Souls". Gain 112 and Abry 115, unfortunately, do not address this issue.

[54] In the *Somnium Scipionis* ("Dream of Scipio") in *De republica* (completed in 51 B.C.), Cic. wrote: "for all who have preserved, assisted, or extended their fatherland, a fixed place has been assigned in the heavens where they may enjoy an everlasting existence, free from care" (*omnibus qui patriam conservaverint, adiuverint, auxerint, certum esse in caelo definitum locum, ubi beati aevo sempiterno fruantur, Rep*. 6.13). The souls of those mortals are said to "consist of those everlasting fires that you call constellations and stars" ([*hominibus*] *animus est ex illis sempiternis ignibus quae sidera et stellas vocatis*, ibid. 6.15).

[55] Cic. *Rep*. 6.16; Manilius 1.758-804; the concept may be traced back to the 4th cent. B.C. Greek philosopher Heraclides of Pontus, a student of Plato: see Gottschalk 98-105, 149-54.

[56] "And it [the Milky Way] passes through the stars of Cassiopeia who is turned upside down" (*transitque inversae per sidera Cassiepiae*, Manilius 1.686).

further incentive for him to embrace that zodiacal sign and make it his own. When viewed in this light, it is not difficult to imagine how the future emperor may privately have welcomed the *sidus Iulium* as a sign of his own "(re)birth" into the Julian gens and, what is more, as the means of achieving the coveted title *Divi filius* ("son of the deified [Julius Caesar]"). Following the games and the fortuitous appearance of the *sidus Iulium* , there could be no doubt that Octavian, a lad not yet nineteen years of age, would have to be taken seriously as a potential successor to Julius Caesar.

APPENDIX I

GRECO-ROMAN SOURCES

Previous collections of sources have included some, but not all of the texts presented here. For instance, de Schodt 388–402, gives 1–6, 9–10, 11A, 20–22, 25–26, 28–31, 33; Pascal 56 n. 6 gives 1–1A, 3–4, 6, 9, 12, 20–22, 24, 29–30, 32; Gardthausen II.1 24–25 n. 26, gives 1–6, 9, 16,18; and Barrett 95–96, gives 1–3, 5–6, 9–10, 12, 33.

We group the texts as follows:

Within each section and subsection, texts are arranged, so far as possible, in chronological order of composition, and for each source we give the year (or range of years) in which the work was written or published. The text of Servius Auctus is indicated by italics, following the convention adopted by G. Thilo (Leipzig 1881–7). The translations are our own and aim at giving the sense of each passage while remaining as faithful as possible to the Greek or Latin. Whenever the word "star" or "comet" occurs, we give in parenthesis the Greek or Latin word that is being translated because the terms vary a great deal in our sources, in part reflecting a debate in July 44 over whether the heavenly body was truly a "comet" (*cometes, sidus crinitum, stella crinita, stella comans*) or rather a new "star" (*sidus, stella, astrum, astron, aster*): see esp. Texts 3 and 6 for evidence of this disagreement and p. 139 for our discussion. We also comment on the Greek or Latin words that are used to give the position of the comet since these terms are subject to more than one interpretation. In this way attention is directed to these key words in our sources, and the reader can see at a glance what lies behind our translation without having to search the entire Greek or Latin text that faces each English version.

I. Sources Mentioning the Comet and Games

A. Games: *ludi Veneris Genetricis*

Text 1: C. Julius Caesar Augustus, *Commentarii de vita sua* fr. 6 (Malcovati)

Ipsis ludorum meorum diebus sidus crinitum per septem dies in regione caeli[a] sub septentrionibus est conspectum. Id oriebatur circa undecimam horam diei clarumque et omnibus e terris conspicuum fuit. Eo sidere significari vulgus credidit Caesaris animam inter deorum immortalium numina receptam, quo nomine id insigne simulacro capitis eius, quod mox in foro consecravimus, adiectum est.

> (a) caeli sub *EFR¹aop:* caeli quae sub *ER²d vett.*:
> caeli quae sub . . . conspectum <est>. *Rackham*

Text 1A: C. Plinius Secundus, *Naturalis Historia* 2.93–94

Cometes in uno totius orbis loco colitur in templo Romae, admodum faustus Divo Augusto iudicatus ab ipso, qui incipiente eo apparuit ludis quos faciebat Veneri Genetrici non multo post obitum patris Caesaris in collegio ab eo instituto. [94] Namque his verbis in . . . gaudium[b] prodit is: '**Ipsis ludorum meorum diebus . . . adiectum est.**' [Text 1] Haec ille in publicum; interiore gaudio sibi illum natum seque in eo nasci interpretatus est. Et, si verum fatemur, salutare id terris fuit.

> (b) in gaudium *MSS: lacuna coni. Mayoff, suppl.* vitae suae memoria
> *vel* commentariis (*Mayoff*), publicum (*Beaujeu*)

I. Sources Mentioning the Comet & Games

A. Games: *ludi Veneris Genetricis*

Text 1: Augustus, *Memoirs* fr. 6 (Malcovati)[1]   [ca. 24 B.C.]
      ap. Pliny, *NH* 2.94 (Text 1A, below)

On the very days of my games [the *ludi Veneris Genetricis*, Text 1A], a comet (*sidus crinitum*) was visible over the course of seven days, in the northern region of the heavens (lit. "under the *septentriones*"[2]). It rose at about the eleventh hour of the day (= approx. 5–6:15 P.M.) and was bright and plainly seen from all lands. The common people believed that this star (*sidus*) signified that the soul of Caesar had been received among the spirits of the immortal gods. On this account, it [a star (*sidus*)] was added as an adornment to the head of the statue of Caesar that I not long afterwards dedicated in the Forum.[3]

Text 1A: Pliny, *Natural History* 2.93–94   [completed A.D. 77]

In only one place in the whole world is a comet (*cometes*) an object of cult, in a temple at Rome. The late emperor Augustus, now deified, judged it to be very propitious to himself, and it appeared at the beginning of his career, at the games that he was celebrating in honor of Venus Genetrix, not long after the death of his father Julius Caesar. Augustus was a member of the board that had been established by Caesar to give the games. [94] With these words he made known his joy in <his *Memoirs*(?)>: **"On the very days . . . . dedicated in the Forum."** [Text 1] These are the sentiments that he expressed publicly. Privately he took pleasure in the view that it [the comet] had come into being for him and that he was coming into being in it. And, if we admit the truth, it was beneficial for the world.

---

[1] Apart from literary allusions (Texts 20-24, 31), this is the earliest extant Greco-Roman text to attest the comet of 44 B.C. The work covered the life of the emperor Augustus through the end of his campaign in Spain against the Cantabri (26-25 B.C.): Suet. *Aug.* 85.1. Scholars assume that it was composed shortly after that campaign, when Augustus returned to Rome in 24: Schanz-Hosius II 11, 398; Bramble 492, Yavetz (1984) 5.

[2] The *septentriones* are the seven bright stars in Ursa Major; the reference may be to the stars familiarly known as the Big Dipper or simply to the "northern quarter of the sky", "the north" (s.v. *OLD* 1a and 2a respectively). We take it in the latter sense: see pp. 86-90 for discussion.

[3] Cf. Suet. *Iul.* 88 (Text 9). According to Dio 45.7.1 (Text 3), this statue was placed in the temple of Venus Genetrix, and so "in the Forum" refers to the new Julian Forum, unless we assume that the statue was housed there only temporarily until it could be moved to the temple of Divus Iulius in the Forum Romanum when that shrine was at last completed and dedicated on 18 Aug. 29 B.C.: so Pesce 403.

Text 2: L. Annaeus Seneca, *Naturales Quaestiones* 7.17.2

[cometes] qui post excessum divi Iulii ludis Veneris Genetricis circa undecimam horam diei emersit.

Text 3: Cassius Dio Cocceianus, *Historiae Romanae* 45.6.4–7.1

[6.4] καὶ μετὰ τοῦτο τὴν πανήγυριν τὴν ἐπὶ τῇ τοῦ 'Αφροδισίου ἐκποιήσει καταδειχθεῖσαν, ἣν ὑποδεξάμενοί τινες ζῶντος ἔτι τοῦ Καίσαρος ἐπιτελέσειν ἐν ὀλιγωρίᾳ, ὥσπερ που καὶ τὴν τῶν Παριλίων ἱπποδρομίαν, ἐποιοῦντο, αὐτὸς ἐπὶ τῇ τοῦ πλήθους θεραπείᾳ, ὡς καὶ προσήκουσαν διὰ τὸ γένος, τοῖς οἰκείοις τέλεσι διέθηκε. [6.5] καὶ τότε μὲν οὔτε τὸν δίφρον τὸν τοῦ Καίσαρος τὸν ἐπίχρυσον οὔτε τὸν στέφανον τὸν διάλιθον ἐς τὸ θέατρον ἐσήγαγεν ὥσπερ ἐψήφισατο, φοβηθεὶς τὸν 'Αντώνιον· [7.1] ἐπεὶ μέντοι ἄστρον τι παρὰ πάσας τὰς ἡμέρας ἐκείνας ἐκ τῆς ἄρκτου πρὸς ἑσπέραν ἐξεφάνη, καὶ αὐτὸ κομήτην τέ τινων καλούντων καὶ προσημαίνειν οἷά που εἴθωθε λεγόντων οἱ πολλοὶ τοῦτο μὲν οὐκ ἐπίστευον, τῷ δὲ Καίσαρι αὐτὸ ὡς καὶ ἀπηθανατισμένῳ καὶ ἐς τὸν τῶν ἄστρων ἀριθμὸν ἐγκατειλεγμένῳ ἀνετίθεσαν, θαρσήσας χαλκοῦν αὐτὸν ἐς τὸ 'Αφροδίσιον, ἀστέρα ὑπὲρ τῆς κεφαλῆς ἔχοντα, ἔστησεν.

Text 3A: Johannes Zonaras, *Epitome Historiarum* 10.13

ἐπεί τις ἀστὴρ τότε ἐφάνη ἐξ ἄρκτου πρὸς ἑσπέραν, ὃν οἱ μὲν κομήτην ἔλεγον προσημαίνοντα οἷά που εἴωθεν, οἱ δὲ τῷ Καίσαρι αὐτὸν ἀντετίθεσαν ὡς τὸν τῶν ἀστέρων ἀριθμὸν ἐγκατειλεγμένῳ, θαρσήσας ὁ πρώην μὲν 'Οκτάβιος, ἤδη δὲ Καῖσαρ κληθείς, μετὰ ταῦτα μέντοι καὶ Αὔγουστος, χάλκεον ἀνδριάντα τοῦ Καίσαρος ἀστέρα φέροντα ὑπὲρ κεφαλῆς εἰς τὸ 'Αφροδίσιον ἔστησεν.

Text 2: Seneca, *Natural Questions* 7.17.2   [ca. A.D. 63]

[The comet of A.D. 60 was not like the one] that rose after the death of the deified Julius, at about the eleventh hour of the day (= approx. 5–6:15 P.M.), at the games in honor of Venus Genetrix.

Text 3: Dio Cassius, *Roman History* 45.6.4–7.1 [completed shortly after A.D. 229]

[6.4] And after this, with a view to winning the favor of the multitude and on the grounds that it was his concern because of his family, Octavian gave at his own expense the festival that had been introduced at the completion of the temple to Venus. Certain persons had undertaken during Caesar's lifetime to discharge the responsibility for giving these games, but they were holding them in slight regard, just as they also neglected the chariot races that were to be given at the *Parilia.* [6.5] On that occasion, because he feared Antony, Octavian did not bring into the theater Caesar's gilded throne and jeweled crown, honors which had been voted to him. [7.1] However, when a certain star (*astron*) appeared throughout all those days in the north[4] towards evening—and although some called it a comet (*cometes*) and said that it portended the usual things, the multitude did not believe this but ascribed it to Julius Caesar, supposing him to have been made immortal and enrolled among the number of the stars—Octavian gained confidence and set up in the temple of Venus a bronze statue of Caesar having a star (*aster*) above its head.[5]

Text 3A: Zonaras, *Epitome of Histories* 10.13 (drawing on Dio 45.7.1)[6]
                                                                    [12th cent. A.D.]

When a star (*aster*) appeared in the north, towards evening, which according to some was a comet (*cometes*) portending the usual things, but others attributed it instead to Caesar, saying that it signified his enrollment among the stars, Octavian took courage—previously that was his name, but at that time he was called Caesar and afterwards Augustus—and set up in the temple of Venus a bronze statue of Caesar having a star (*aster*) above its head.

---

[4] Lit. "from (the direction of) the bear", *arktos* being the name of the constellation Ursa Major and also a word that can mean simply "the north", (s.v. LSJ I.2 and 3 respectively); so πρὸς ἄρκτον ("towards the north") giving the location of the comet of 426 B.C. (Aristot. *Mete.* 1.6.343b) and *a septentrione* (lit. "from the seven bright stars in Ursa Major", the Latin equivalent of Dio's ἐκ τῆς ἄρκτου) giving the location where the Claudian comet (of A.D. 54?) was first seen (Sen. *Nat. qu.* 7.29.3). We take ἐκ τῆς ἄρκτου in this latter sense: see pp. 86-90 for discussion.

[5] See n. 3 on Text 1.

[6] Although this source does not attest the name of the games, we include it here, numbering it 3A, since it has no independent evidentiary value, but is wholly dependent on Dio (Text 3).

Text 4: Iulius Obsequens, *Prodigiorum Liber* 68

Ludis Veneris Genetricis, quos pro collegio fecit, stella hora undecima crinita sub septentrionis sidere exorta convertit omnium oculos. Quod sidus quia ludis Veneris apparuit, divo Iulio insigne capitis consecrari placuit.

B. Games: *ludi Veneris Genetricis et funebres*

Text 5: Servius Honoratus (*Danielis*), *Commentarii* in Vergilii *Aen.* 8.681

APERITUR VERTICE SIDUS: *apparet* sidus in vertice, hoc est super galeam. nam ex quo tempore per diem stella visa est, dum sacrificaretur Veneri Genetrici et ludi funebres Caesari exhiberentur, *per triduum stella apparuit in septentrione.* quod sidus *Caesaris* putatum est Augusto persuadente: *nam ideo Augustus omnibus statuis, quas divinitati Caesaris statuit, hanc stellam adiecit. Ipse vero Augustus* in honorem patris stellam in galea coepit habere depictam.

C. Games: *ludi funebres*

Text 6: Servius Honoratus (*Danielis*), *Commentarii* in Vergilii *Ecl.* 9.47

ECCE DIONAEI PROCESSIT CAESARIS ASTRUM: cum Augustus Caesar ludos funebres patri celebraret, die medio stella apparuit. Ille eam esse confirmavit parentis *sui:* unde sunt versus isti compositi. 'Dionaei' autem longe repetitum est, a matre Veneris Diona. Sane 'astrum' graece dixit: nam stellam debuit dicere. *Baebius Macer circa horam octavam stellam amplissimam, quasi lemniscis, radiis coronatam,[(a)] ortam dicit. Quam quidam ad inlustrandam gloriam Caesaris iuvenis pertinere existimabant, ipse animam patris sui esse*

(a) quasi lemniscis, radiis *Thilo:* quasi lemnicas cactis *cod.:*
quasi lemniscatis *Daniel:* quasi radiis lemniscatis *Heinsius:*
quasi lemniscis *Masvicius*

Text 4: Julius Obsequens, *Book of Prodigies* 68 (for the year 44 B.C.)

[4th cent. A.D.?]

At the games in honor of Venus Genetrix, which he gave on behalf of the board, a comet (*stella crinita*) rose in the north (lit. "under the constellation *septentrio*"[7]) at the eleventh hour (= approx. 5–6:15 P.M.) and caught the eye of everyone. The decision was made to dedicate this star (*sidus*) to the deified Julius, as an adornment of his head, because it appeared at the games in honor of Venus.

B. Games: *ludi Veneris Genetricis et funebres*

Text 5: Servius on Virgil, *Aeneid* 8.681   [4th cent. A.D.]

"At the top of his head, [his father's] star (*sidus*) is revealed": a star (*sidus*) appears on the top of his head, that is on the upper part of his helmet. For Augustus himself began to have a star (*stella*) represented on his helmet in honor of his father (Caesar) from the time when a star (*stella*) was seen during the daytime while sacrifices were being performed to Venus Genetrix and funeral games were being presented for Caesar. *It appeared in the north*[8] *for a three-day period,*[9] and Augustus encouraged the view that it was *Caesar's star* (*sidus*). *And for this reason Augustus added this star* (stella) *to all the statues which he set up in honor of Caesar's divinity.*[10]

C. Games: *ludi funebres*

Text 6: Servius on Virgil, *Eclogue* 9.47 [Text 20]   [4th cent. A.D.]

"Behold, the star (*astrum*) of Dionean Caesar came forth": when Augustus Caesar was holding funeral games in honor of his father, a star (*stella*) appeared at midday. He asserted positively that it was the star (*stella*) of his own father Caesar: hence these verses were composed. "Dionean" makes remote allusion to Dione, the mother of Venus. The poet plainly uses a Greek word '*astrum*' for "star" in place of '*stella*'. *Baebius Macer*[11] *says that a very large star* (stella) *rose*

---

[7] See n. 2 on Text 1.

[8] Lit. *in septentrione*, see n. 2 on Text 1.

[9] For a possible explanation of this variant tradition (Augustus said it was visible for 7 days, Text 1), see p. 130.

[10] Our other sources speak of only one statue which Octavian set up in the Forum (see n. 3 on Text 1).

[11] Most likely a contemporary observer: see p. 84.61.

voluit eique in Capitolio statuam, super caput auream stellam habentem, posuit: inscriptum in basi fuit 'Caesari emitheo'. Sed Vulcanius[(b)] *aruspex in contione dixit cometem esse, qui significaret exitum noni saeculi et ingressum decimi; sed quod invitis diis secreta rerum pronuntiaret, statim se esse moriturum: et nondum finita oratione, in ipsa contione concidit. Hoc etiam Augustus in libro secundo De memoria vitae suae complexus est.*

(b) Vulcanius *cod.* (*cf.Schulze, Eigennamen 377*): Vulcatius *Masvicius*

Text 7: Servius Honoratus (*Danielis*), *Commentarii* in Vergilii *Aen.* 1.287

IMPERIUM OCEANO, FAMAM QUI TERMINET ASTRIS: aut ad laudem dictum est, aut certe secundum historiam. Re vera enim et Britannos qui in oceano sunt vicit, et post mortem eius cum ludi funebres ab Augusto eius adoptivo filio darentur, stella medio die visa est, unde est 'ecce Dionaei processit Caesaris astrum' (*Ec.* 9.47).

Text 8: Servius Honoratus (Danielis), *Commentarii* in Vergilii *Aen.* 6.790

CAELI VENTURA SUB AXEM: nam cum Augustus patri Caesari ludos funebres exhiberet, stella per diem apparuit, quam persuasione Augusti Caesaris esse populus credidit. Hinc est 'ecce Dionaei processit Caesaris astrum' (*Ec.* 9.47). 'Sub axem' ergo, id est ad divinos honores.

at about the eighth hour (= approx. 1:15–2:30 P.M.),[12] one that was surrounded with rays, like streamers on a garland. *Some persons thought that it (the star) had to do with furthering the glory of the young Caesar* (Octavian), *but he himself wanted it to be the soul of his father and dedicated to him on the Capitoline a statue which had a golden star (*stella*) above its head. On the base was the inscription "to the demigod Caesar."*[13] *But the haruspex Vulcanius stated in a public meeting that it was a comet (*cometes*) which indicated the end of the ninth and the beginning of the tenth age. He also stated that he would die on the spot, because he was announcing, against the will of the gods, matters that lay hidden. And he collapsed without finishing his speech, in the very assembly. Augustus too included this information in book two of his* Memoirs.[14]

Text 7: Servius on Virgil, *Aeneid* 1.287   [4th cent. A.D.]

"(Caesar) who will mark the limits of his sway by the ocean, the limits of his fame by the stars (*astra*)": this statement though intended as praise, is strictly in keeping with historical fact. For in reality Caesar conquered the Britons who live on an island in the ocean, and after his death, when funeral games were being given in his honor by his adopted son Augustus, a star (*stella*) was seen at midday,[15] whence the words "Behold, the star (*astrum*) of Dionean Caesar came forth" [Text 20].

Text 8: Servius on Virgil, *Aeneid* 6.790   [4th cent. A.D.]

"Destined to come beneath the pole of heaven": For when Augustus was celebrating funeral games in honor of his father Caesar, a star (*stella*) appeared during the daytime, and Augustus convinced the people to believe it was Caesar's [star]. Hence the line: "Behold, the star (*astrum*) of Dionean Caesar came forth" [Text 20]. "Beneath the pole" then refers to achieving divine honors.

---

[12] For a possible explanation of this variant tradition (Augustus said it became visible at the 11th hour, Text 1), see p. 130-31.

[13] As Weinstock 41 n. 6 has observed, this source confuses the statue erected to Caesar on the Capitoline in 46 B.C. (Dio 43.14.6) with the statue dedicated by Octavian in the Forum (see n. 3 on Text 1).

[14] As we can tell from the Latin *etiam Augustus* ("Augustus also"), Servius' account of Vulcanius and his prophecy was drawn chiefly from Baebius Macer, not Augustus, who is cited only to say that he too treated the episode in his *Memoirs*. Augustus is likely to have reinterpreted Vulcanius' prophecy, applying it to the dawning of a new age of peace and prosperity that was to follow the conclusion of the civil wars in 31 B.C.: see pp. 142-43.

[15] See pp. 130-31 for a possible explanation of the discrepancy between this account and the Augustan tradition (Text 1, cf. 2, 4, 9) that the comet appeared at the 11th hour (= approx. 5:15-6 P.M.), while another eyewitness(?), Baebius Macer (Text 6), gave the 8th hour (= approx. 1:15-2:30 P.M.).

D. Games: Name not specified

Text 9: C. Suetonius Tranquillus, *Vita Divi Iulii* 88

Periit sexto et quinquagensimo aetatis anno atque in deorum numerum relatus est, non ore modo decernentium, sed et persuasione vulgi. Siquidem ludis, quos primo consecratos[a] ei heres Augustus edebat, stella crinita per septem continuos dies fulsit exoriens circa undecimam horam, creditumque est animam esse Caesaris in caelum recepti; et hac de causa simulacro eius in vertice additur stella.

> (a) primo consecratos (-tis *L1 d) lectio archetypi:* primos (*om.* consecratos) *G1 g* (s *in G erasa m. 2):* primo consecrato *ed. Basil. 1533:* primos consecrato *ed. Basil. 1546:* primos consecratos *Stephanus*

II. Sources Mentioning the Comet Only

Text 10: L.(?) Mestrius Plutarchus, *Vita Caesaris* 69.3

τῶν δὲ θείων ὅ τε μέγας κομήτης (ἐφάνη γὰρ ἐπὶ νύκτας ἑπτὰ μετὰ τὴν Καίσαρος σφαγὴν διαπρεπης, εἶτα ἠφανίσθη) καὶ τὸ περὶ τὸν ἥλιον ἀμαύρωμα τῆς αὐγῆς.

Text 11: Cassius Dio Cocceianus, *Historiae Romanae* 45.17.4
Text 11A: Xiphilinus, *Epitome Dionis*

λαμπὰς ἀπ' ἀνίσχοντος ἡλίου πρὸς δυσμὰς διέδραμε, καί τις ἀστὴρ καινὸς ἐπὶ πολλὰς ἡμέρας ὤφθη.

Text 12: Iulius Obsequens, *Prodigiorum Liber* 68

Fax caelo ad occidentem visa ferri. Stella per dies septem insignis arsit.

D. Games: Name not specified

Text 9: Suetonius, *Life of Julius Caesar* 88
[completed shortly before A.D. 121/2]
    He died in the fifty-sixth year of his life and was assigned to the company of the gods not only by the text of a decree but also by the conviction of the common people. For at the games which Caesar's heir Augustus established in his honor and celebrated for the first time, a comet (*stella crinita*) shone throughout seven days in a row, rising at about the eleventh hour (= approx. 5–6:15 P.M.), and it was believed that it was the soul of Caesar who had been taken up into heaven. For this reason a star (*stella*) is added to his statue above his head.[16]

II. Sources Mentioning the Comet Only

Text 10: Plutarch, *Life of Caesar* 69.3   [ca. A.D. 110–120]

    Of the divinely sent signs there was a large comet (*cometes*)—it appeared prominently for seven nights after the murder of Caesar, then vanished from sight—and the dimming about the Sun of its rays.

Text 11: Dio Cassius, *Roman History* 45.17.4
[completed shortly after A.D. 229]
Text 11A: Epitome of Dio by Xiphilinus, ed. Dindorf, vol. 5 p. 40.14–16
[11th cent. A.D.]

    A torch (*lampas*) ran across the sky from east to west, and a new star (*aster*) was seen for many days.

Text 12: Julius Obsequens, *Book of Prodigies* 68 (for the year 44 B.C.)
[4th cent. A.D.?]
    In the sky a torch (*fax*) was seen to be carried along towards the west. A star (*stella*) shone conspicuously for seven days.

---

[16] See n. 3 on Text 1.

Text 13: Rufius Festus Avienius ap. Serv. (*Daniels*) Aen. 10.272

Est etiam alter cometes, qui vere cometes appellatur; nam comis hinc inde cingitur. Hic blandus esse dicitur. Qui si orientem attenderit, laetas res ipsi parti significat; si meridiem, Africae aut Aegypto gaudia; si occidentem inspexerit, terra Italia voti sui compos erit. Hic dicitur apparuisse eo tempore quo est Augustus sortitus imperium; tunc denique gaudia omnibus gentibus futura sunt nuntiata.

III. Texts Mentioning the Games Only

A. Games: *ludi Veneris Genetricis*

Text 14: Nicolaus Damascenus, *Vita Caesaris Augusti* fr. 130 28.108

Καῖσαρ δ' οὐδὲν ὀρρωδῶν ἐκ τοῦ μεγαλόφρονος θέας ἐποίει ἐνστάσης ἑορτῆς, ἣν ὁ πατὴρ αὐτοῦ κατεστήσατο 'Αφροδίτῃ. καὶ αὖθις παρελθὼν σὺν πλείοσιν ἔτι καὶ φίλοις παρεκάλει 'Αντώνιον συγχωρῆσαι τὸν δίφρον μετὰ τοῦ στεφάνου τίθεσθαι τῷ πατρί. ὁ δ' ὅμοια ἠπείλησεν, εἰ μὴ τούτων ἀποστὰς ἡσυχίαν ἄγοι. καὶ ὃς ἀπῄει καὶ οὐδὲν ἠναντιοῦτο κωλύοντος τοῦ ὑπάτου· εἰσιόντα γε μὴν αὐτὸν εἰς τὸ θέατρον ἐκρότει ὁ δῆμος εὖ μάλα καὶ οἱ πατρικοὶ στρατιῶται ἠχθημένοι διότι τὰς πατρῴους ἀνανεούμενος τιμὰς διεκωλύθη, ἄλλους τε ἐπ' ἄλλοις κρότους ἐδίδουν παρ' ὅλην τὴν θέαν ἐπισημαινόμενοι.

Text 15: Appianus, *Bella Civilia* 3.28.107

(ὁ 'Αντώνιος) ἐκώλυσε δὲ (sc. αὐτὸν προτιθέναι τόν τε χρύσεον θρόνον καὶ στέφανον) καὶ ἐν ταῖς ἑξῆς θέαις ἔτι παραλογώτερον, ἃς αὐτὸς ὁ Καῖσαρ ἐτέλει, ἀνακειμένας ἐκ τοῦ πατρὸς 'Αφροδίτῃ Γενετείρᾳ, ὅτε περ αὐτῇ καὶ τὸν νεὼν ὁ πατὴρ τὸν ἐν ἀγορᾷ ἅμα αὐτῇ ἀγορᾷ ἀνετίθει.

Text 13: Avienius ap. Servius on Virgil, *Aeneid* 10.272 [4th cent. A.D.]

There is furthermore another comet which is truly called "comet" [cometes = "having flowing hair"], *for it is surrounded on all sides by hair.[17] This one is said to be gentle. If it watches over the east, it portends joyous happenings for that part of the world. If it beholds the south, then joyous events are in store for Africa and Egypt. If it looks upon the west, the land of Italy will achieve its fondest wishes. This type (of comet) is said to have appeared at the time when Augustus obtained power by lot. Then finally the joys that were in store for all nations were announced.*

III. Texts Mentioning the Games Only

A. Games: *ludi Veneris Genetricis*

Text 14: Nicolaus of Damascus, *Life of Augustus* fr. 130. 28.108

[last quarter of 1st cent. B.C.]

As a result of his high-mindedness Octavian was not at all daunted, but he presented the games belonging to the festival that was at hand which his father had established to Venus. And with still more friends he went again to Antony and called upon him to permit the chair together with the crown to be set up in honor of his father Caesar. Antony made threats similar to those before if he did not abandon these requests and remain quiet. Octavian went off and did not put up opposition since the consul was forbidding him. Yet the common people gave him a big round of applause when he entered the theater, and his father's soldiers were annoyed because he had been prevented from renewing his father's honors, and they signified their approval of Octavian by giving him round after round of applause throughout the whole festival.

Text 15: Appian, *Civil Wars* 3.28.107 [mid 2nd cent. A.D.]

With less reason did Antony prevent Octavian [from setting up the gilded chair and crown] also at the games that came next which Octavian himself discharged. These games had been established by his father Caesar in honor of Venus Genetrix at the time when Caesar dedicated the temple to Venus in the new Forum Iulium together with the forum itself.

---

[17] Compare the description given by Baebius Macer, who may have been a contemporary observer: "a very large star . . . surrounded by rays, like streamers on a garland" (Text 6).

B. Games: *ludi Victoriae Caesaris*

Text 16: C. Matius, [Cicero] *Epistulae ad familiares* 11.28.6

"At ludos quos Caesaris victoriae Caesar adulescens fecit curavi." At id ad privatum officium, non ad statum rei publicae pertinet; quod tamen munus et hominis amicissimi memoriae atque honoribus praestare etiam mortui debui et optimae spei adulescenti ac dignissimo Caesare petenti negare non potui.

Text 17: C. Suetonius Tranquillus, *Vita* Augusti 10.11

Ludos autem victoriae Caesaris non audentibus facere quibus optigerat id munus, ipse edidit.

C. Games: Name not specified

Text 18: M. Tullius Cicero, *Epistulae ad Atticum* 15.2.3

De Octavi contione idem sentio quod tu, ludorumque eius apparatus et Matius ac Postumus mihi procuratores non placent; Saserna conlega dignus. Sed isti omnes, quem ad modum sentis, non minus otium timent quam nos arma.

Text 19: M. Tullius Cicero, *Epistulae ad familiares* 11.27.7

Defensio autem est duplex: alia sunt quae liquido negare soleam, ut de isto ipso suffragio, alia quae defendam a te pie fieri et humane, ut de curatione ludorum.

B. Games: *ludi Victoriae Caesaris*

Text 16: Gaius Matius[18] to Cicero in *Cicero's Letters to his Friends* (*Fam.*)
11.28.6   [Aug.–Oct. 44 B.C.]
    "But I superintended the games that young Caesar [Octavian] celebrated in honor of [Julius] Caesar's victory." Yet this act concerns a private obligation (*officium*), and has nothing to do with the condition of the commonwealth. Even so, I owed the discharge of this duty (*munus*)[19] to the memory and honor of a very dear, departed friend. Also I could not refuse Octavian's request since he is a youth of the highest promise and most worthy of [the name] Caesar.

Text 17: Suetonius, *Life of Augustus* 10.1 [completed shortly before A.D. 121/2]

    He himself gave the games in honor of Caesar's victory since those upon whom this duty had fallen did not venture to give the games.

C. Games: Name not specified

Text 18: Cicero, *Letters to Atticus* (*Att.*) 15.2.3   [18 May 44 B.C.]

    Concerning Octavian's speech to the people[20] I share your views. I dislike the preparations for his games and his agents, Matius[21] and Postumus.[22] Saserna[23] is a worthy colleague. All those persons, as you well know, fear peace and stability just as much as we fear war and upheaval.

Text 19: Cicero to Gaius Matius in *Cicero's Letters to his Friends* (*Fam.*) 11.27.7
                                                          [Aug.–Oct. 44 B.C.]
    My defense of your conduct is twofold. There are some charges that I am accustomed to deny unequivocally, such as the one concerning your vote at the assembly. Other acts I defend on the grounds that you carried them out from a sense of obligation and human decency, such as in the instance of your superintending Octavian's games.

---

[18] A learned man, friend of Cicero and Caesar; like Cicero's friend Atticus, he never stood for political office but remained a member of the equestrian order.

[19] On the meaning of *munus* in this context, see p. 49 and p. 49.29.

[20] For the date, see p. 1.1.

[21] See n. 18.

[22] Gaius Rabirius (Curtius) Postumus, successfully defended by Cicero in late 54/early 53 B.C., in a trial for extortion; praetor 48 B.C.(?): *MRR* III 181. He was a financier of shady character.

[23] One of three brothers; possibly Gaius (or Publius?) Hostilius Saserna, tribune of the plebs in 44 B.C. (*MRR* III 104).

IV. Literary Allusions to the Comet

A. Star

Text 20: P. Vergilius Maro, *Ecloga* 9.46–49

> Daphni, quid antiquos signorum suspicis ortus?
> ecce Dionaei processit Caesaris astrum,
> astrum quo segetes gauderent frugibus et quo
> duceret apricis in collibus uva colorem.

Text 21: P. Vergilius Maro, *Aeneidos* Liber 8.681

> . . . . . patriumque aperitur vertice sidus.

Text 22: Q. Horatius Flaccus, *Carmina* 1.12.46–48

> . . . . micat inter omnis
> Iulium sidus velut inter ignis
>   luna minores.

Text 23: Sex. Propertius, *Elegiae* 3.18.33–34

> . . . . et qua
> Caesar, ab humana cessit in astra via.

Text 24: Sex. Propertius, *Elegiae* 4.6.59

> at pater Idalio miratur Caesar ab astro.

IV. Literary Allusions to the Comet

A. Star

Text 20: Virgil, *Eclogue* 9.46–49   [composed ca. 42 B.C.]

Why do you pay attention to the risings of the old constellations, Daphnis? Behold, the star *(astrum)* of Dionean Caesar came forth as a sign for the crops to take delight in their heads of grain and the grape to take on its color on the sunny hills.

Text 21: Virgil, *Aeneid* 8.681   [26–19 B.C.]

At the top of his [Augustus'] head, his father's star *(sidus)* is revealed.

Text 22: Horace, *Odes* 1.12.46–48   [composed ca. 24 B.C.]

The Julian star *(sidus)*[24] outshines all others just as the Moon outshines the lesser lights.

Text 23: Propertius, *Elegies* 3.18.33–34   [composed late 23 B.C.]

He [Marcellus] withdrew from the path of mortals to the stars *(astra)* by the same route that Caesar took.

Text 24: Propertius, *Elegies* 4.6.59   [last quarter of 1st cent. B.C.]

But his father Caesar marvels as he looks down from the Idalian star *(astrum)*.

---

[24] Although "*Iulium sidus*" is doubtless intended principally as a reference to Augustus (see Weinstock 378-79), this passage is included here because the expression may also be reasonably interpreted as alluding indirectly to the comet (so Nisbet-Hubbard ad loc., pp. 162-63) and has been frequently treated in modern scholarship as synonymous with Caesar's comet—e.g., in works by Gardthausen, Pesce, de Schodt, Scott, and Wagenvoort.

Text 25: Valerius Maximus, *Facta et Dicta Memorabilia* 1 praefatio

si excellentissimi vates a numine aliquo principia traxerunt, mea parvitas eo iustius ad favorem tuum decucurrerit, quo cetera divinitas opinione colligitur: tua praesenti fide paterno avitoque sideri par videtur, quorum eximio fulgore multum caerimoniis nostris inclitae claritatis accessit.

Text 26: Valerius Maximus, *Facta et Dicta Memorabilia* 3.2.19

Sed ut armorum togaeque prius, nunc etiam siderum clarum decus, Divum Iulium, certissimam verae virtutis effigiem, repraesentemus.

Text 27: Valerius Maximus, *Facta et Dicta Memorabilia* 6.9.15

C. Caesar . . . clarissimum mundi sidus . . .

Text 28: Silius Italicus, *Punica* 13.862–64

> . . . . ille deum gens,
> stelligerum attolens apicem, Troianus Iulo
> Caesar avo;

B. Comet: propitious

Text 29: P. Ovidius Naso, *Metamorphoses* 15.746–50

> Caesar in urbe sua deus est; quem Marte togaque
> praecipuum non bella magis finita triumphis
> resque domi gestae properataque gloria rerum
> in sidus vertere novum stellamque comantem,
> quam sua progenies.   750

Text 25: Valerius Maximus, *Memorable Deeds and Words* I, preface (addressed to Tiberius, the Roman Emperor)　　　　　　　[published soon after A.D. 31]

If the most distinguished bards have begun their works with an address to some divine power, all the more rightly shall I, a humble author, have recourse to your favor in that the divinity of the other powers is amassed by mere opinion, whereas yours, as a result of its manifest reliability, appears equal to the star (*sidus*) of your father and grandfather. As a result of their outstanding brightness, much renowned brilliance has been added to our religious solemnities.

Text 26: Valerius Maximus, *Memorable Deeds and Words* 3.2.19

[published soon after A.D. 31]

But let us portray the deified Caesar, previously foremost in distinction in war and peace and even now foremost of the stars (*sidera*), a most certain representative of true virtue.

Text 27: Valerius Maximus, *Memorable Deeds and Words* 6.9.15

[published soon after A.D. 31]

Gaius Caesar . . . brightest star (*sidus*) of the universe . . .

Text 28: Silius Italicus, *Punica* 13.862–64　　　　[last decade of 1st cent. A.D.]

That one, a descendant of the gods, the top of whose head is crowned with a star (*stelli-gerum*), is Caesar, descended from Trojan Julus.

B. Comet: propitious

Text 29: Ovid, *Metamorphoses* 15.746–50　　　　　　　　　[ca. A.D. 8]

Caesar in his own city is a god. An outstanding figure in war and peace, he was turned into a new star (*sidus*), a comet (*stella comans*), as much by his descendant (i.e., Augustus[25]), as by the wars that he brought to a conclusion with triumphs, by his civil accomplishments and by his mortal glory that was quickly won.

---

[25] Augustus secured Caesar's recognition as a god by a formal decree of the Senate and law in early 42 (see p. 56.51).

Text 30: P. Ovidius Naso, *Metamorphoses* 15.840–50

> hanc animam interea caeso de corpore raptam   840
> fac iubar, ut semper Capitolia nostra forumque
> divus ab excelsa prospectet Iulius aede!"
> Vix ea fatus erat, media cum sede senatus
> constitit alma Venus nulli cernenda suique
> Caesaris eripuit membris neque in aera solvi   845
> passa recentem animam caelestibus intulit astris.
> dumque tulit, lumen capere atque ignescere sensit
> emisitque sinu: luna volat altius illa,
> flammiferumque trahens spatioso limite crinem
> stella micat . . .   850

C. Comet: baleful

Text 31: P. Vergilius Maro, *Georgicon*   Liber 1.487–88

> Non alias caelo ceciderunt plura sereno
> fulgura nec diri totiens arsere cometae.

Text 32: Albius Tibullus, *Elegiae* 2.5.71

> haec fore dixerunt belli mala signa cometen

Text 33: T. Calpurnius Siculus, *Ecloga* 1.77–83

> Cernitis ut puro nox iam vicesima caelo
> fulgeat et placida radiantem luce cometem
> proferat? Ut liquidum niteat sine vulnere plenus?
> Numquid utrumque polum, sicut solet, igne cruento   80
> spargit et ardenti scintillat sanguine lampas?
> At quondam non talis erat, cum Caesare rapto
> indixit miseris fatalia civibus arma.

Text 30: Ovid, *Metamorphoses* 15.840–50   [ca. A.D. 8]

" . . . Meanwhile, snatch this soul from its slain body and turn it into a brilliant light so that deified Julius [Caesar] may always look down from his lofty shrine upon our Capitoline and Forum."

He (Jupiter) had scarcely finished uttering these words when kindly Venus stood in the midst of the Senate, invisible to all. She snatched from his limbs her descendant Caesar's newly released soul and did not allow it to dissolve into thin air. She bore it to the heavenly stars, and while she carried it, she perceived that it was taking light and blazing into flame. She released it from her bosom, and it flew higher than the Moon; it shone as a star (*stella*), trailing its fiery hair along an ample track.

C. Comet: baleful

Text 31: Virgil, *Georgic* 1.487–88 (among omens foretelling civil war after the death of Caesar)   [36–29 B.C.]

At no other time did more lightning bolts fall from a clear sky; never did fearsome comets (*cometae*) so often blaze.

Text 32: Tibullus, *Elegies* 2.5.71 (Sibylline prochecy of civil war after death of Caesar)   [composed shortly before the poet's death in 19 B.C.]

These (sc. the Sibyls) said that a comet (*cometes*) would be the evil sign of war.

Text 33: Calpurnius Siculus, *Eclogue* 1.77–83 (comparing the gentle light of the comet of A.D. 60(?) with the ominous, blood-red comet of 44)
                                   [ca. mid 1st cent. A.D.]

Do you behold how now for the twentieth time the night gleams with a clear sky and brings forth a comet (*cometes*) beaming with gentle light? and how it shines serenely, full and without harm? Surely it does not drench either hemisphere with bloody fire, as comets usually do, and the torch (*lampas*) does not glitter with glowing blood? But not like this one was the one (comet) in times past when it announced the destined war for the wretched citizens of Rome after Caesar had been snatched from them.

APPENDIX II

THE DATE OF THE DEDICATION
OF THE TEMPLE OF VENUS GENETRIX IN 46 B.C.

Scholars generally state that the temple of Venus Genetrix was dedicated in 46 B.C. on 26 September, citing the imperial *fasti* as evidence (e.g., Wissowa 292; Koch, *RE* 865). The reform of the calendar in 46, however, introduces an element of uncertainty as to the precise day on which the dedication took place. To be sure, the date of the anniversary attested in the imperial *fasti*, *a. d. VI Kal. Oct.*, corresponds to 26 September, but in the pre-Julian calendar, which was still in use in 46, the month of September had only 29 days, and so the sixth day before the Kalends (1st) of October, by inclusive reckoning, was 25 September. To make our position clear, we should state that in our view the Julian date given by the *fasti* for the temple's *dies natalis* (*a. d. VI Kal. Oct.*) stands for *a. d. V Kal. Oct.* in the pre-Julian calendar. (This is also the view of Degrassi 514). That is, the interval between the Ides (13th) of September and the dedication of the temple was kept constant by adjusting the numerical designation of the day relative to the Kalends of the following month so as to compensate for the lengthening of September from from 29 to 30 days in the Julian calendar. In both calendars, the dedication (in 46) and its anniversary (in 45 and subsequent years, after the adoption of the Julian calendar) fell on 26 September.

We support this conclusion with the following observations:

(1) If the Romans had failed to compensate for the longer months in the Julian calendar by adjusting the names of days on which festivals and other anniversaries fell, three of the seven major sets of annual games would have been lengthened by one day since the last day of those games fell after the Ides in months to which a day had been added (the *Ceriales* in April, *Romani* in September and *Plebeii* in November). The dates of these games attested in the early imperial *fasti*, after the introduction of the Julian calendar, are:

Ceriales: prid. Id. Apr. to *a. d. XIII Kal. Mai.*= 12-19 April (12-18 in pre-Julian calendar)
Romani: prid. Non. Sept. to *a. d. XIII Kal. Oct.* = 4-19 Sept. (4-18 Sept. in pre-Julian calendar)
Plebeii: prid. Non. Nov. to *a. d. XV Kal. Dec.* = 4-17 Nov. (4-16 Nov. in pre-Julian calendar)

If the numerical designation of the day on which each of these festivals ended remained unchanged when the conversion was made from the pre-Julian to the Julian calendar, then one day was automatically added to each of these *ludi*, as shown by the pre-Julian equivalents of these dates that are given in parentheses. The silence of our sources, which fail to comment on such a windfall for the theater-going public (to say nothing of the financial burden for the poor aediles), argues against an automatic extension of these three festivals as a result of Caesar's new calendar. Presumably, therefore, when the new calendar was introduced the length of the *ludi* was kept constant by renaming the day on which each festival ended: the last day of the *Ceriales*, for instance, being changed from *a. d. XII Kal Mai.* (= 19 April) in the pre-Julian calendar to *a. d. XIII Kal. Mai.* (= 19 April) in the Julian calendar.[1] If so, and if the length of the *ludi Romani* was likewise kept constant by renaming the day in the new calendar on which it concluded, then presumably the interval between that day and the anniversary of Venus' temple will have been kept constant by renaming the *dies natalis* of the temple: *a. d. V Kal. Oct.* (26 September pre-Julian) becoming *a. d. VI Kal. Oct.* (26 September Julian), the date attested in the imperial *fasti*.

(2) According to Macrobius (*Sat.* 1.14.11), the Romans sought to preserve the *ordo* (established order) of the *feriae* (religious festivals) in each month by changing the name, where necessary, of certain days that were *festus* or *feriatus* (given over to feasts or festivals) so that they would continue to fall in the new calendar on the 15th of the month (*tertio ab Idibus die*). That is, holidays that had been celebrated on *a. d. XVI Kal.* in the pre-Julian calendar were shifted to *a. d. XVII* or *XVIII Kal.* in months that gained one or two days respectively. Macrobius' statement is confirmed by the evidence of the republican *Fasti*

---

[1] A similar adjustment in date to maintain the overall length of a festival when going from the pre-Julian to the Julian calendar may possibly be seen in at least one other instance. In an unpublished paper on the rites of the *argei* (delivered in December 1995 at the Annual Meeting of the APA), Gary Forsythe has argued that the two different dates given by our sources for this rite (15 May, according to Dion. Hal. *Ant. Rom.* 1.38.3, and 14 May, according to Ovid, *Fasti* 5.621-22) may best be explained by assuming that when April was increased from 29 to 30 days in the Julian calendar, the date of the rites were made to fall one day earlier (moved from 15 to 14 May) so as to preserve the overall length of the rites which occupied precisely 60 days, from the stationing of the *argei* in their chapels on 16 March, to their being tossed into the Tiber in May (on the 15th in the pre-Julian calendar that was known to Dionysius' sources, and on the 14th in the imperial calendar known to Ovid): 16 days Mar. + 29 days Apr. (pre-Julian) + 15 days May = 60 days, compared with 16 days Mar. + 30 days Apr. (Julian) + 14 days May = 60 days.

Antiates Maiores, compared with the imperial *fasti*, for such *feriae* as the *Carmentalia* 15 Jan. (*XVI Kal.* pre-Julian, *XVIII Kal.* Julian), *Fordicidia* 15 Apr. (*XVI Kal.* pre-Julian, *XVII Kal.* Julian), and *Consualia* 15 Dec. (*XVI Kal.* pre-Julian and *XVIII Kal.* Julian).

(3) In personal communication (2 Jan. 1996), Gary Forsythe has pointed out to us that the date of the *Robigalia* remained 25 April, necessitating an a change from *a. d. VI Kal. Mai.* in the pre-Julian calendar (as attested by the *Fasti Ant. Mai.*) to *a. d. VII Kal. Mai.* in the Julian calendar (as attested by the imperial *fasti*, *CIL* I² 316).

APPENDIX III

THE DATE OF THE *LUDI VENERIS GENETRICIS* IN 46 B.C.

Taylor, *Divinity* 63 is no doubt correct to conclude that in 46 the games to Venus Genetrix followed directly upon the celebration of Caesar's four triumphs (over Gaul, Egypt, Pontus and Africa). This view is to be preferred to Weinstock's: triumphs in August (p. 76) and games a few weeks later (p. 79). Weinstock's scheme is theoretically possible since Caesar reached Rome on 25 July in 46 B.C. (*BAfr*. 98), but according to his reconstruction, the triumphs and games would no longer be so closely connected as they are in our sources. Gelzer 284 states that the four triumphs and games occupied 20 September to 1 October (eleven days) no doubt basing this assumption merely on the fact that the *ludi Victoriae Caesaris* later occupied the eleven days 20-30 July.

Most likely Caesar's decision to dedicate his new temple to Venus Genetrix after holding a series of triumphs influenced Octavian to do likewise in 29 B.C. (Dio 51.21.5-7, 22.4-9): a triple triumph on three successive days (13,14, 15 August as we know from the *fasti*), followed by the dedication of his temple to Divus Iulius (18 August, attested by the *fasti*). (If Octavian was following the precedent set by Julius Caesar in 46 B.C., it is possible to understand why the *aedes Divi Iuli* was dedicated in August rather than July, which might have seemed more appropriate because not only had July been named in Caesar's honor, but it was the month in which both his birthday and *ludi Victoriae Caesaris* were celebrated.) Caesar's triumphs in 46 occupied one day each (Dio 43.19.1), with intervals in between (Suet. *Iul*. 37.1 *interiectis diebus*). This takes us up to 26 September if we assume that the first triumph was held on 20 September, allowing one day (19 September) to separate the *ludi Romani* from the series of triumphs (on 20, 22, 24, 26 September) on the analogy of the one day (11 April) that served to separate the *Megalenses* from the *Ceriales* (cf. the two days, 2-3 September, that separated Pompey's *ludi votivi* from the *Romani* in 70, *Verr*. 31). The last day of the triumph will have coincided, therefore, with the dedication of the temple (so Taylor, *Divinity* 63 and Koch, *RE* 865 assume, while Rawson 453 puts the dedication "on the day after his last triumph"). Caesar's decision to cause the final day of his triumph to fall on the day he had chosen for the dedication of his new temple to the ancestress of the Julian family may be compared with Pompey's arrangements in 61: a triumph lasting two days (Plut. *Pomp.* 45), the second day being his birthday (Pliny, *NH* 7.98, 37.13). As further evidence that the dedication of the temple most likely fell on the last day of Caesar's triumph in 46, we have Dio's statement (43.22.1) that after the banquet (δεῖπνον: presumably the *epulum* of Suet. *Iul.* 38.2; cf. Plut. *Caes.* 55.1, 22,000 *triclinia*) on the evening of the final triumph a large crowd, with elephants bearing torches, escorted Caesar home from his new forum where the temple of Venus Genetrix was the showpiece. The games will have commenced on the following day. (It was customary for games to follow, not precede the dedication of a temple: e.g., the temple of Juno and the temple of Diana in 179, Livy 40.52.3; *aedes Fortunae* in 173, Livy 42.10.5; *aedes Divi Iuli* in 29, Dio 42.10.5.) To judge from the descriptions of their variety and splendor (Nic. Dam. 9.19; Suet. *Iul.* 39; Plut. *Caes.* 55.2; Dio 43.22.3-23.6) , the games must have run well into October, and in 45 the festival will have been repeated, commencing on 26 September and extending into early October.

APPENDIX IV

THE DESCRIPTION OF CAESAR
AS VICTORY'S "NEIGHBOR"[1]

Scholars generally assume that Cicero alludes to Caesar as Victory's "neighbor" (*vicinus*) in *Att.* 13.44.1 because he was thinking of the circus procession (*pompa*) in which the statues of Caesar and Victory were being carried next to each other.[2] This interpretation works equally well, of course, for any set of public games, the *ludi Apollinares* (6-13 July) as well as the *ludi Victoriae Caesaris* (20-30 July), but deeper reflection should cause us to realize that *vicinus* is an odd choice for expressing this idea. If the word *vicinus* was intended merely to call attention to the fact that statues of Caesar and Victory were carried "side by side" (so Shackleton Bailey), might not one have expected Caesar to be called Victory's *comes* ("companion") or *sodalis* ("comrade", "crony"), rather than *vicinus* ("neighbor")? Indeed, the standard interpretation of *vicinus* in this passage presumes a sense of the word (= "temporary proximity"), which upon closer inspection *vicinus* turns out not to have. The noun *vicinus*, even when used adjectivally with the meaning "situated close to each other" (*OLD* s.v. 3a), appears to be used elsewhere almost exclusively of more permanent relationships of proximity than the one envisaged on the occasion of the *pompa*.[3] We have been unable to find the word *vicinus* employed anywhere else in Cicero with the meaning that it is generally assumed to have in *Att.* 13.44.1.

Possibly another interpretation of *vicinus* in *Att.* 13.44.1 is worth considering, one that takes into account the way Cicero uses this word in some other letters. For instance, in a letter to Atticus, Cicero describes the goddess Salus as "your neighbor

---

[1] We thank Russell T. Scott and our colleague Elizabeth Gebhard for reading and commenting on an earlier draft of this section of our study and for sharing with us their views on some of the relevant features of Roman topography. The responsibility for the interpretation offered here is, of course, entirely our own. We also thank Prof. Scott and Prof. James Russell for drawing to our attention some of the recent scholarly literature on these issues.
[2] So Taylor 230, Weinstock 185, cf. 91, and Shackleton Bailey ad loc.
[3] Shackleton Bailey in oral communication has expressed agreement with this observation.

Salus" (*tuae vicinae Salutis, Att.* 4.1.4) because the temple of Salus was near Atticus' house on the Quirinal Hill. Likewise, in Vettius' testimony of 59 B.C., Cicero was styled the "neighbor of the consul [Caesar]" (*vicinus consulis, Att.* 2.24.3) because Cicero's house on the north-east slope of the Palatine[4] overlooked the *domus publica* (NE of the temple of Vesta) in the Forum, Caesar's official residence in Rome from 63 until his death.[5] Finally, Caesar is twice jocularly called Atticus' *vicinus* (*Att.* 12.45.2, 12.48.1) because Caesar's statue had recently been placed in the temple of Quirinus on the Quirinal, where Atticus had his home.[6]

In the light of these other passages, we believe that the reference to Caesar as Victory's *vicinus* in *Att.* 13.44 is most likely also based on a feature of Rome's topography. To judge from the use of *vicinus* in the four letters just cited, we should expect Caesar to be identified as Victory's "neighbor" because there was a shrine of Victory not far from the *domus publica*, where Caesar lived. One shrine, and possibly two, may in fact meet this requirement. The first of these is the shrine of Vica Pota, an early Italic goddess, who was identified by the Romans with the goddess Victory.[7] In two different sources this goddess is actually called "Victoria", implying that the assimilation was so complete that the names were interchangeable.[8] Her shrine, we know, lay at the foot of the Velia, on the site of

---

[4] On Cicero's house, see Richardson (1) 123 and Allen (2) 134-43.

[5] Caesar moved from a house in the Subura to the *domus publica* on the *via Sacra* after his election as *pontifex maximus* (chief priest) in 63 (Suet. *Iul.* 46; cf. Pliny *NH* 19.23 describing the awning with which Caesar temporarily covered the Forum during the games in 46, stretching "from his own house" [*ab domo sua*]). The house stood on the north perimeter of the precinct of Vesta, opposite the so-called temple of Romulus, at the beginning of the slope of the *via Sacra* (*primore clivo*). It was a substantial residence, whose remains reveal that modifications were made in the late republic, probably by Caesar so that the house would better suit his needs as his principal residence when in Rome. See Coarelli (1986) 21-23 and Richardson (1) 133-34. We thank Russell Scott for discussing with us the recent excavations of this site.

[6] Gelzer 307 n. 2, Richardson (1) 122, and a few others have concluded on the basis of these references that the new house that was to be built for Caesar at public expense (Dio 43.44.6) was to be located on the Quirinal. (See Weinstock 171 for a refutation of this view.) Wissowa 155 n. 7, Shackleton Bailey (ad loc.), and Weinstock (loc. cit.) more plausibly connect Cicero's remark with Caesar's statue in the temple of Quirinus (attested by Dio 43.45.3). This must have been Cicero's meaning, as revealed by the barbed comment in *Att.* 12.45.2 that he preferred Caesar to be housed with the god Quirinus rather than Salus (goddess of Safety).

[7] See Ogilvie on Livy 2.7.12 and Weinstock, *RE* A8.2 (1958) 2014-15, who accepts Cicero's view (*Leg.* 2.28) that Vica Pota's name was related to *vincendi atque potiundi* ("conquering and gaining mastery").

[8] In Asconius p. 13 (Clark) 15, "below the Velia, where there is now the shrine of Victory" (*sub Veliis ubi nunc aedis Victoriae est*) and in a speech falsely attributed to Cicero (*Oratio priusqam in exilium iret* 24), "Jupiter Stator . . . whose temple Romulus established together with that of Victory at the foot of the Palatine, after the conquest of the Sabines" (*Iuppiter Stator . . . cuius templum a Romulo victis Sabinis in Palatii radice cum Victoria est collocatum*). For a discussion of the evidence furnished by this latter passage, see Ziolkowsi (1989) 135.

the house that had once belonged to Publius Valerius Publicola.[9] It was situated therefore, presumably a little to the east of the *domus publica*, towards the beginning of the *clivus* leading up the Velia.[10]

The second shrine that could also account for the description of Caesar as Victory's *vicinus* is the Palatine temple of Victory. Unfortunately the exact location of that temple is a matter of speculation. One clue to its location is furnished by a passage that attests a *Clivus Victoriae* on the Palatine Hill.[11] Presumably the temple of Victory stood at the top of this *clivus*, since we know that these sloping roads typically took their names from the temples that stood at their termini (e.g., the *Clivus Capitolinus* and *Clivus Salutis*), but unfortunately even the route followed by this *clivus* on the Palatine is today no longer agreed upon. Up until recently the *Clivus Victoriae* was identified with the remains of a sloping pathway that began on the north-western side of the Palatine and curved sharply eastward at the northern corner of the hill, just above the imperial *via Nova*.[12] This identification, based upon a fragment of the Severan *Forma Urbis Romae* (frag. 42) that clearly preserves the words "*Cli>vus Victoria<e*" and what were once thought to be the remains of the *Horrea Agrippiana* at the foot of the north-western slope of the Palatine, flanked by the *Vicus Tuscus*, has now been shown probably to belong to the Caelian, rather than the Palatine Hill.[13] We can no longer say for certain, therefore, that Cicero, Clodius and Metellus lived on the *Clivus Victoriae*, just above the *via Nova*.[14]

Recent speculation has encouraged us to look for the temple of Victory on the south-west corner of the hill, at the head of the *Scalae Caci*, to the east of the temple of the Magna Mater,[15] and excavations directed by Patrizio Pensabene

---

[9] Livy 2.7.12 "below the Velia . . . at the bottom of the slope" (*infra Veliam . . . in infimo clivo*) and Ascon. (above, p. 186.8). See Coarelli (1986) 80-82.

[10] See Ziolkowski (1992) 171-72.

[11] Festus p. 318 Lindsay: said to be *infimus* ("lowest") in the vicinity of the *Porta Romanula*, which lay at the north corner of the Palatine. See Coarelli (1986) 228-34.

[12] So, for instance, Platner-Ashby 126 and Nash 257.

[13] As demonstrated by Carettoni et al. 109-11. For a reconsideration of the evidence in the light of a previously unpublished inscription (in the Museo Nazionale, Rome), which mentions a *Vicus Victoriae*, see Panciera 238-41. Coarelli (1980) 38-39 and 123 continues to show the *Clivus Victoriae* and *Horrea Agrippiana* where they were once thought to be depicted by frag. 42. We thank Russell Scott for calling Panciera's article to our attention.

[14] As, for instance, Richardson (1) 91, 123, 124 does. For the suggestion that the path above the precinct of Vesta, on the NE slope of the Palatine which runs under the substructures of the domus Tiberiana and which was once thought to be the *Clivus Victoriae*, may in fact be the republican *via Nova*, see Morganti-Tomei 560-63.

[15] This was one of two spots tentatively suggested by Castagnoli 185-86. Wiseman 35-52 has argued at length for the site adjoining the temple of the Magna Mater. Most of Wiseman's conclusions have recently been accepted by Ziolkowski (1992) 172-74, and they have led Rehak 182-84 to argue that a tetrastyle Ionic temple shown on a fragmentary relief in the Capitoline Museum may possibly represent the Palatine temple of Victory.

during the past decade have uncovered the remains of a temple which could be the long sought after *aedes Victoriae*.[16] The grounds for making this identification, however, are not terribly compelling: (1) a passage in Dionysius of Halicarnassus (*Ant. Rom.* 1.33.1) crediting Evander's Arcadians with building a temple to Nike (Victory) "on the highest point of the crest of the hill" (ἐπὶ δὲ τῇ κορυφῇ τοῦ λόφου); (2) inscriptions attesting the temple of Victory (a republican *cippus* [boundary stone] and fragment of an Augustan architrave[17]) whose exact find-spots are not recorded with precision but seem to have been just south of the modern church of S. Teodoro, on the north-west side of the Palatine;[18] and (3) a wall-painting from the house of Marcus Fabius Secundus in Pompeii, which shows what may be the temple of Victory in the context of the story of the procreation of Romulus and Remus. All of this evidence is quite tenuous. Wiseman who discusses the painting at some length has to admit that the shrine, which he identifies with the primitive Arcadian temple of Nike/Victory, is made to overlook what appears to be the temple of Vesta (next to the site of Caesar's home in the Forum). This he interprets as the painter's compression of the "north-western side of the Palatine in order to bring the temple of Vesta into his scene" (Wiseman 37). However, as we have noted, the evidence of *Att.* 13.44.1 may suggest that the temple of Victory was perhaps on the north-eastern slope, overlooking the *domus publica* where Caesar lived, and in that case the two temples, of Victory and Vesta, will have been, in fact, about as they are shown in the wall-painting.[19]

---

[16] See Pensabene 54-58, reporting the discovery of the remains of a podium measuring 33 x 20 meters.

[17] *CIL* VI.3733 (= VI.31059, I² 805, *ILLRP* 284) and VI.31060. To these may be added an inscription preserved on the base of a votive offering ([. . .] *L. l(ibertus) donum solvit Victoriae* "Lucius, a freedman, paid for this gift to Victory"), recently discovered by Pensabene in the *domus Augustana* to the NE of the temple of the Magna Mater (Schippa 283-95).

[18] See Wiseman 37-40 and Fig. 1.

[19] For its possible location, see Tab. A in Lugli, marked "*aedes Victoriae?*".

APPENDIX V

PARALLELS BETWEEN AUGUSTUS' *MEMOIRS* AND LATER SOURCES

(Sources are arranged from left to right in the order of their composition)

Conflicting accounts are enclosed within square brackets.

AUGUSTUS Text 1	MACER ap. Text 6	SENECA Text 2	PLUTARCH Text 10	SUETONIUS Text 9	DIO Text 3	SERVIUS Texts 5, 6, 7, 8	OBSEQUENS Texts 4, 12
per septem dies			ἐπὶ νύκτας ἑπτά	per septem continuos dies	[παρὰ πάσας τὰς ἡμέρας ἐκείνας]	[per triduum](5)	per dies septem insignis arsit(12)
in regione caeli sub septentrionibus					ἐκ τῆς ἄρκτου	in septentrione(5)	sub septentrionis sidere(4)
oriebatur circa undecimam horam diei	[circa horam octavam] ortam	circa undecimam horam diei emersit		exoriens circa undecimam horam	πρὸς ἑσπέραν	per diem (5, 8) [die medio] (6, 7)	hora undecima(4) exorta(4)
clarumque et omnibus e terris conspicuum fuit							convertit omnium oculos(4)
Eo sidere significari volgus credidit Caesaris animam inter deorum immortalium numina receptam				persuasione volgi . . . creditumque est animam esse Caesaris in caelum recepti	ἀπηθανατισμένῳ καὶ ἐς τὸν τῶν ἄστρων ἀριθμὸν ἐγκατειλεγμένῳ		
id insigne simulacro capitis eius, quod mox in foro consecravimus, adiectum est				hac de causa simulacro eius in vertice additur stella	χαλκοῦν αὐτὸν ἐς τὸ Ἀφροδίσιον, ἀστέρα ὑπὲρ τῆς κεφαλῆς ἔχοντα, ἔστησεν	Augustus omnibus statuis, quas divinitati Caesaris statuit, hanc stellam adiecit(5) statuam, super caput auream stellam habentem, posuit(6)	Quod sidus . . . divo Iulio insigne capitis consecrari placuit(4)

APPENDIX VI

WAS THE *SIDUS IULIUM* PERHAPS
A NOVA OR SUPERNOVA?

Given the facts that the *sidus Iulium* (1) was seen for only seven days, (2) was daylight visible, and (3) had a bright, star-like appearance, there is a chance that it was a nova (or supernova), rather than a comet, but the likelihood is rather remote. A supernova is a star that undergoes a cataclysmic explosion at the end of its evolution. Daylight-visible supernovae have been recorded: for instance, the new star of A.D. 1054, which was daylight visible for 23 days and produced the Crab Nebula (Clark-Stephenson 150). However, the maximum brightness of a supernova declines relatively slowly, extending over a period of weeks. The seven-day duration of the Roman observation would tend to rule out a supernova.

A nova, on the other hand, is a white dwarf star having a normal star as its companion. Matter from the atmosphere of the normal star falls onto the surface of the white dwarf. When the density of the accreted material becomes high enough, it undergoes a thermonuclear explosion. No daylight-visible nova has been recorded (Payne-Gaposchkin). The probability of a seeing a nova in daylight for seven days in any given year can be estimated as about 1/3000.[1]

---

[1] Daylight visibility lasting only seven days is an unusually rapid decay for a nova. The fastest recorded decay rate is 0.3 magnitudes per day (Clark-Stephenson). At this rate, if it reached magnitude -4 on the seventh day, the maximum apparent brightness would have been -4 - 0.3 x7 = -6.1 . Fast novae have absolute magnitudes around -8.5 (Payne-Gaposchkin). The absolute magnitude m_{ab} is related to the apparent magnitude m_{ap} by the equation $m_{ab} + 5 \log(r/10) = m_{ap}$, where r is the distance from the Earth in parsecs. This gives a value for r of 38 parsecs. If we take r less than 45 parsecs, following Clark-Stephenson, we would expect such novae to occur at a rate of $0.1*(r/300)^3 = .00034$ / yr, or about once every 3000 years.

APPENDIX VII

WAS THERE A SOLAR ECLIPSE VISIBLE
FROM ROME IN 44 B.C.?

Two solar eclipses are known to have occurred in 44, in April and October, but neither was visible from Rome: see Oppolzer 112-13 with chart 56, for 18 April and 11 October, corrected to early 12 October in ephemeris time by Mucke-Meeus 260 & charts 734.[1] Yet according to Pliny (*NH* 2.98) the period following Caesar's murder was marked by "portentous and long-lasting eclipses [?] of the Sun" (*prodigiosi et longiores solis defectus*). Possibly the author of *De viris illustribus* (78.10) refers to this same tradition when he states that on the day of Caesar's funeral (i.e., ca. 20 March: Becht 84-85) "the Sun hid its orb" (*cuius corpore pro rostris posito sol orbem suum celasse dicitur*). Servius (on *Georg.* 1.466) reports a *solis defectus* that supposedly took place on 14 May 44, lasting "from the sixth hour, all the way to night" (*ab hora sexta usque ad noctem*), = ca. 10:50 to 19:00 for that time of year and latitude (see Drumann-Groebe 773, "Stundentafel"). Eight hours is clearly too long a period for an eclipse to last, and we happen to have a letter written by Cicero on 14 May at Puteoli (*Att.* 14.22) in which there is no mention of any unusual phenomena, solar or otherwise.

The only way to make sense of Servius' eight hours is to understand *defectus* not as an "eclipse" but rather as a sudden "darkening" of the Sun. In other words, the meaning of the noun *defectus* in the passages from Pliny and Servius approaches that of an expression employing the adj. *defectus* ("reduced", "enfeebled") that we find in Tibullus (2.5.75) where the Sun is described as "reduced in brilliance" throughout the "cloudy year" 44 : *Solem defectum lumine . . . nubilus annus*. Petau (X.65 II.147-50) deserves credit for noting in the 17th century that *defectus* in the passage from Pliny cited above should not be taken literally in the sense of "eclipses." Concerning the passage in Servius describing the *solis defectus* on 14 May, Petau writes: "who ever heard of a true solar eclipse lasting for 6 hours?" (*quis enim veram solis defectionem audiit usquam sex horas durasse?*).

---

[1] We thank the Adler Planetarium for calling our attention to Mucke-Meeus.

As we argue in chapter five (pp. 99-107), the probable explanation of the sudden darkening reported by Servius is that it was a product of the eruption of Etna, but modern scholars have generally clung to the notion that an eclipse was indeed observed from Italy in 44. They merely note that Servius' date should be corrected from May to November: so most recently R. F. Thomas on *Georg.* 1.466. However, this is a mistaken notion, and it may be traced back ultimately to Wunderlich's revision of Heyne's commentary on Virgil. To Heyne's note on *Georg.* 1.466, Wunderlich added the assertion that a solar eclipse took place in November, citing as his authority *Sammlung astronomischer Tafeln* II (Berlin: G. Decker, 1776) 122. From Wunderlich, this misinformation has made its way into most subsequent commentaries (e.g., those by Forbiger, Conington-Nettleship, Page, and Thomas), the notable exceptions being Haverfield, who in the 5th revised ed. (1898) of Conington-Nettleship cautioned that no eclipse had occurred in November 44, and Mynors, who was content to dismiss the notice in Servius without referring to the "corrected" date in November. The tradition that there was an solar eclipse visible from Italy in November 44 is by now so well established in the modern scholarly literature on Virgil that we shall probably never be able to quash this mistaken belief entirely.

APPENDIX VIII

THE DETERMINATION OF A PARABOLIC ORBIT FROM TWO SIGHTINGS PLUS PERIHELION DISTANCE

We denote quantities pertaining to the May sighting by the subscript "1", to the July sighting by "2". For $i = 1, 2$, let: $\mathbf{u}_i$ denote the unit vectors along the Earth to Comet lines, $\mathbf{R}_i$ the Sun to Earth vectors, $\mathbf{r}_i$ the Sun to Comet vectors, θ_i the true anomalies, r_i the Earth-Comet distances, and t_i the observation times. Let t_0 denote the time of perihelion and q the perihelion distance. From the dynamics of cometary orbits, for $i = 1, 2$:

$$r_i = \frac{2q}{1 + \cos(\theta_i)} = q\left[1 + \tan^2\left(\frac{\theta_i}{2}\right)\right] \quad . \tag{1}$$

The vectors $\mathbf{r}_i$ are related to the Earth's position vectors $\mathbf{R}_i$ and the sighting directions $\mathbf{u}_i$ by

$$\mathbf{r}_i = \rho_i \mathbf{u}_i + \mathbf{R}_i \quad . \tag{2}$$

With $T_i = t_i - t_0$, let

$$V_i = \frac{3\,k\,T_i}{2\sqrt{2}\,q^{3/2}} \quad , \tag{3}$$

where k is Gauss' constant. Let $\Theta = \tan(\theta/2)$. The times of observation, the perihelion date and distance determine the true anomalies θ through

$$\Theta = \left[V + \sqrt{V^2 + 1}\right]^{1/3} - \frac{1}{\left[V + \sqrt{V^2 + 1}\right]^{1/3}} \quad , \tag{4}$$

where we have omitted the subscript i. From Eq. (2) we have

$$\rho_i = -\mathbf{u}_i \cdot \mathbf{R}_i \pm \sqrt{(\mathbf{u}_i \cdot \mathbf{R}_i)^2 - R_i^2 + r_i^2} \quad . \tag{5}$$

195

Only the positive root is physical for the July sighting, but both roots are permitted for the May sighting. The positive root gives a high inclination orbit, the negative one gives an orbit with low inclination. For fixed q, t_1 and t_2, this equation, along with Eqs. (1), (3) and (4), gives the distances ρ_i as functions of the date of perihelion t_0.

The essence of the orbit fitting algorithm is to consider the vector inner product $\mathbf{r_1 \cdot r_2}$. This is on the one hand expressible, using Eq. (1), as

$$\mathbf{r_1 \cdot r_2} = r_1 r_2 \cos\left(\theta_2 - \theta_1\right) \tag{6}$$

$$= q^2\left[1 - \Theta_1^2 - \Theta_2^2 + \Theta_1^2\Theta_2^2 + 4\Theta_1\Theta_2\right] \quad,$$

which, with fixed q, t_1 and t_2, is a function, G, of the date of perihelion, t_0. On the other hand, we also have

$$\mathbf{r_1 \cdot r_2} = \rho_1\rho_2\mathbf{u_1 \cdot u_2} + \rho_1\mathbf{u_1 \cdot R_2} + \rho_2\mathbf{u_2 \cdot R_1} + \mathbf{R_1 \cdot R_2} \quad, \tag{7}$$

which is another function of t_0, denoted by F. The equation, G-F = 0, is solved for t_0 using a standard method of successive approximations. Eqs. (4) and (1) then give the distances r_i, Eq. (5) gives the distances ρ_i, and the vectors $\mathbf{r_i}$ are constructed using Eq. (2).

Consider the set of unit vectors; $\mathbf{n}$ perpendicular to the plane of the orbit, $\mathbf{e}$ perpendicular to the plane of the ecliptic, $\mathbf{a}$ pointing to the first point of Aries, $\mathbf{p}$ pointing from the Sun to the position of perihelion, $\mathbf{N}$ pointing from the Sun to the ascending node; plus the unit vector:

$$\mathbf{b} = \mathbf{n} \times \mathbf{r_1} \quad . \tag{8}$$

The vector $\mathbf{n}$ is obtained from:

$$\mathbf{n} = \frac{\mathbf{r_2} \times \mathbf{r_1}}{|\mathbf{r_2} \times \mathbf{r_1}|} \quad, \tag{9}$$

the vector $\mathbf{p}$ is:

$$\mathbf{p} = \cos\left(\theta_1\right)\frac{\mathbf{r_1}}{r_1} + \sin\left(\theta_1\right)\mathbf{b} \quad, \tag{10}$$

and **N** is:

$$N = \frac{nxe}{|nxe|} \quad . \tag{11}$$

Then the longitude of the ascending node, Ω, is:

$$\Omega = \arccos(\mathbf{N} \cdot \mathbf{a}) \quad . \tag{12}$$

The inclination of the orbit , I, comes from:

$$I = 180 - \arccos(\mathbf{n} \cdot \mathbf{e}) \quad , \tag{13}$$

and the argument of perihelion, ω, is given by:

$$\omega = \text{sign}(\mathbf{p} \cdot \mathbf{e})\arccos(\mathbf{N} \cdot \mathbf{p}) \quad . \tag{14}$$

BIBLIOGRAPHY

Abry   Abry, Josèphe-Henriette. "Auguste: La Balance et le
                        Capricorne." *REL* 66 (1988): 103–21.

Alföldi (1930)   Alföldi, A. "Der neue Weltherrscher der vierten Ekloge
                        Vergils." *Hermes* 65 (1930): 369–84.

Alföldi (1953)   ———. *Studien über Caesars Monarchie.* Lund:
                        Gleerup, 1953.

Alföldi (1973)   ———. "La divinisation de César dans la politique
                        d'Antoine et d'Octavien entre 44 et 40 avant J.C."
                        Revue numismatique 6 ser. 15 (1973): 99–128.
                        Reprint in *Caesariana.* Bonn: Bahelt, 1984: 229–68.

Alföldi (1976)   ———. *Oktavians Aufsteig zur Macht.* Bonn: Habelt,
                        1976.

Allard et al.   Allard, P., J. Carbonnelle, D. Dajlevic, J. Le Bronec, P.
                        Morel, M. Robe, J. Maurenas, R. Faivre-Pierret, D.
                        Martin, J. Sabroux, and P. Zettwoog. "Eruptive and
                        diffuse emissions of CO_2 from Mount Etna." *Nature*
                        351 (1991): 387–91.

Allen (1)   Allen, Richard. *Star Names, their Lore and Meaning.*
                        1899. Reprint, New York: Dover, 1963.

Allen (2)   Allen, W., Jr. "The location of Cicero's house on the
                        Palatine Hill." *CJ* 35 (1939): 134–43.

AGU Report   American Geophysical Union. *Volcanism and Climate
                        Change.* Special Report. Washington, D.C.: AGU,
                        1992.

Baldet   Baldet, M. Fernand. *Liste Générale des Comètes de
                        l'Origine a 1948,* extrait de *l'Annuaire du Bureau
                        des Longitudes pour l'an 1950.* Paris: Gauthier-
                        Villars, 1950.

Bardon   Bardon, Henry. *La Littérature Latine Inconnue.* Vol. 2.
                        Paris: C. Klincksieck, 1956.

Barrett   Barrett, A. A. "Observations of comets in Greek and
                       Roman sources before A.D. 410." *Journal of the
                       Royal Astronomical Society of Canada* 72 (1978):
                       81–106.

Barton   Barton, Tamsyn. "Augustus and Capricorn:
                       Astrological polyvalency and imperial rhetoric."
                       JRS 85 (1995): 33–51.

de Bary   de Bary, W. T. *Sources of Chinese Tradition.* New
                       York: Columbia Univ. Press, 1960.

Beaujeu   Beaujeu, Jean. *Cicéron, Correspondance* VIII. Budé ed.
                       Paris: Les Belles Lettres, 1983.

Becht   Becht, Erich. "Regeste über die Zeit von Cäsars
                       Ermordung bis zum Umschwung in der Politik des
                       Antonius." Freiburg Diss., 1911.

Beck   Beck, B. J. Mansvelt. *The Treatises of Later Han.*
                       Leiden: Brill, 1990.

Bellemore   Bellemore, Jane. *Nicolaus of Damascus, Life of
                       Augustus.* Bristol: Bristol Classical Press, 1984.

Bernoulli   Bernoulli, Johann. *Römische Ikonographie* I. 1882.
                       Reprint, Hildesheim: Olms, 1969.

Bicknell   Bicknell, Peter. "Blue suns, the son of heaven, and the
                       chronology of the volcanic veil of the 40s B.C."
                       Ancient History Bulletin 7 (1993): 2–11.

Boll   Boll, F. "Beiträge zur Überlieferungsgeschichte der
                       griechischen Astrologie und Astronomie." *SBAW* I
                       (1899): 77–140.

Bömer   Bömer, F. "Über die Himmelserscheinung nach dem
                       Tode Caesars." *BJ* 152 (1952): 27–40.

Bouché-Leclercq   Bouché-Leclercq, A. *L'Astrologie Grecque.* 1899.
                       Reprint, Brussels: Culture et Civilisation, 1963.

Bowersock   Bowersock, G. W. "The pontificate of Augustus." In
                       Between Republic and Empire, edited by K.
                       Raaflaub and M. Toher. Berkeley: Univ. of
                       California Press, 1990: 380–94.

Boyce   Boyce, Aline A. "A new Augustan aureus of 17 B.C."
                       In *Festal and Dated Coins of the Roman Empire:
                       Four Papers.* Numismatic Notes and Monographs
                       no. 153. New York: American Numismatic Society,
                       1965: 1–11.

Bramble   Bramble, J. C. "The age of Augustus: Minor figures" in
                        the *Cambridge History of Classical Literature* II.
                        Cambridge: Cambridge Univ. Press, 1982: 467–94.

Brind'Amour   Brind'Amour, Pierre. *Le Calendrier Romain*. Ottawa:
                        Univ. of Ottawa Press, 1983.

Brunt-Moore   Brunt, P. A., and J. M. Moore, *Res Gestae Divi Augusti*.
                        Oxford: Oxford Univ. Press, 1967.

Butler-Cary   Butler, H. E., and M. Cary. *Suetonius, Divus Julius*.
                        Oxford: Oxford Univ. Press, 1927.

Carettoni et al.   Carettoni, G., A. Colini, L. Cozza, and G. Gatti. *La
                        Pianta Marmorea di Roma Antica: Forma Urbis
                        Romae*. Rome, 1960.

Casson   Casson, Lionel. *Travel in the Ancient World*. Toronto:
                        Hakkert, 1974.

Castagnoli   Castagnoli, F. "Note sulla topographie del Palatino e
                        del Foro Romano." *Arch. Class.* 16 (1964): 173–99.

Chambers   Chambers, George. *Handbook of Descriptive and
                        Practical Astronomy*. I⁴. Oxford: Oxford Univ.
                        Press, 1889.

Ch'en   Ch'en Yüan. *Erh-shih-shih shuo-jun piao*. 1962, rev.
                        ed. Reprint, Peking: Chung-hua shu-chü ch'u-pan,
                        1978.

Clark-Stephenson   Clark, David H. and F. Richard Stephenson. *The
                        Historical Supernovae*. Oxford: Pergamon, 1977.

Coarelli (1980)   Coarelli, Filippo. *Roma*. Bari: Guide Archeologiche
                        Laterza, 1980.

Coarelli (1986)   ———. *Il Foro Romano: Periodo Arcaico*. 2nd ed.
                        Rome: Quasar, 1986.

Coleman   Coleman, Robert. *Virgil, Eclogues*. Cambridge:
                        Cambridge Univ. Press, 1977.

Conington   Conington, John. *P. Vergili Maronis Opera*. Vol. I. 4th
                        ed. rev. by Henry Nettleship. London: Whittaker,
                        1881. 5th ed. rev. by F. Haverfield. London: G. Bell,
                        1898.

Cramer   Cramer, F. *Astrology in Roman Law and Politics*.
                        Philadelphia: American Philosophical Society, 1954.

Crawford   Crawford, M. *Roman Republican Coinage*. 2 vols.
                        Cambridge: Cambridge Univ. Press, 1974.

Cullen

Cullen, C. "Motivations for scientific change in ancient China: Emperor Wu and the grand inception astronomical reforms of 104 B.C." *Journal for the History of Astronomy* 24 (1993): 185–203.

Degrassi

Degrassi, A. *Fasti anni Numani et Iuliani, Inscr. Ital.* XIII. 2. Rome: La Libreria dello Stato, 1963.

Denniston

Denniston, J. D. *Cicero, Philippics I & II.* Oxford: Oxford Univ. Press, 1926.

Drew

Drew, D. L. *The Allegory of the Aeneid.* Oxford: Blackwell, 1927.

Drumann-Groebe

Drumann, W., and P. Groebe. *Geschichte Roms* III². 1906. Reprint, Hildesheim: Olms, 1964.

Dubs

Dubs, H. *The History of the Former Han Dynasty.* 3 vols. Baltimore: Waverly Press, 1938–55.

Duffett-Smith

Duffett-Smith, Peter. *Practical Astronomy with your Calculator.* 2nd ed. Cambridge: Cambridge Univ. Press, 1983.

Ehrenwirth

Ehrenwirth, Ursula. "Kritisch-chronologische Untersuchungen für die Zeit vom 1. Juni bis zum 9. Oktober 44 v. Chr." Munich Diss., 1971.

Fitzpatrick

Fitzpatrick, Philip M. *Principles of Celestial Mechanics.* New York: Academic Press, 1970.

Forbiger

Forbiger, Albertus. *P. Virgilii Maronis Opera.* Vol. I³. Leipzig: Hinrichs, 1852.

Forsythe

Forsythe, P. Y. "In the wake of Etna, 44 B.C." *Classical Antiquity* 7 (1988): 49–57.

Gagé

Gagé, Jean. *Res Gestae Divi Augusti.* 1935. 3rd revised ed. Paris: Les Belles Lettres, 1977.

Gain

Gain, D. B. *The Aratus ascribed to Germanicus Caesar.* London: Athlone, 1976.

Gardiner

Gardiner, K. H. J. *The Early History of Korea.* Canberra: Centre of Oriental Studies and the Australian National Univ. Press, 1969.

Gardthausen

Gardthausen, Viktor. *Augustus und seine Zeit.* 2 vols. in 6 parts. Leipzig: Teubner, 1891–1904.

Garnsey

Garnsey, Peter. *Famine and Food Supply in the Greco-Roman World.* Cambridge: Cambridge Univ. Press, 1988.

Gelzer — Gelzer, Matthias. *Caesar, Politician and Statesman.* Translated by Peter Needham. Cambridge, Mass.: Harvard Univ. Press, 1968.

Getty — Getty, R. J. "Liber et alma Ceres." *Phoenix* 5 (1951): 96–107.

Giard — Giard, Jean-Baptiste. *Catalogue des monnaies de l'empire Romain* I. Paris: Bibliothèque Nationale, 1976.

Gibbon — Gibbon, Edward. *The History of the Decline and Fall of the Roman Empire.* Ed. H. H. Milman. 6 vols. Boston: Phillips, Sampson, 1850.

Goold — Goold, G. P. *Manilius, Astronomica.* Loeb ed. Cambridge, Mass.: Harvard Univ. Press, 1977.

Gottschalk — Gottschalk, H. B. *Heraclides of Pontus.* Oxford: Oxford Univ. Press, 1980.

Gruter — Gruter, Ian. *Inscriptiones Antiquae totius orbis Romani.* Rev. ed. 2 vols. Amsterdam: Halma, 1707.

Gundel — Gundel, Wilhelm. "Kometen." *RE* 11.1 (1921): 1143–93.

Gundel (1926) — ———. "Textkritische und exegetische Bemerkungen zu Manilius." *Philologus* 81 (1926): 309–38.

Gurval — Gurval, Robert. *Actium and Augustus: The politics and emotions of civil war.* Ann Arbor: Univ. of Michigan Press, 1995.

Habel — Habel, Edwin. "Ludi publici." *RE* suppl. 5 (1931): 608–30.

Hahn — Hahn, I. "Zur Interpretation der Vulcatius-Prophetie." *Acta antiqua et archaeologica Szeged* 16 (1968): 239–46.

Halley — Halley, Edmund. *A Synopsis of the Astronomy of Comets.* An English translation of a rev. and expanded ed. of the 1705 *Synopsis Astronomiae Cometicae* printed as an appendix in David Gregory's *The Elements of Astronomy, Physical and Geometrical.* London, 1715: vol. II, 881–905.

Halm-Laubmann — Halm, Karl, and G. Laubmann. *Ciceros ausgewählte Reden.* 10th ed. Berlin: Weidmann, 1899.

Hammer — Hammer, C. U. "Traces of Icelandic eruptions in the Greenland ice sheet." *Jökull* 34 (1984): 51–65.

Hammer et al.   Hammer, C. U., H. B. Clausen, and W. Dansgaard.
                       "Greenland ice sheet evidence of post-glacial
                       volcanism and its climatic impact." *Nature* 288
                       (1980): 230–35.

Harris   Harris, W. *Rome in Etruria and Umbria.* Oxford:
                       Oxford Univ. Press, 1971.

Hasegawa   Hasegawa, Ichiro. "Catalogue of ancient and naked-eye
                       comets." *Vistas in Astonomy* 24 (1982): 59–102.

Haulet et al.   Haulet, R., P. Zettwoog, and J. Sabroux. "Sulphur
                       dioxide discharge from Mount Etna." *Nature* 268
                       (1977): 715–17.

Hazzard   Hazzard, R. A. "Theos Epiphanes: Crisis and
                       response." *Harvard Theological Review* 88 (1995):
                       415–36.

Heinze   Heinze, Richard. *Virgils epische Technik.* 3rd ed.
                       Leipzig: Teubner, 1928.

Henry   Henry, James. *Aeneidea.* Vol. 3. Dublin: Williams and
                       Norgate, 1889.

Herron   Herron, Michael. "Impurity sources of F^-, Cl^-, NO_3^-,
                       and SO_4^{2-} in Greenland and Antarctic precipitation."
                       Journal of Geophysical Research 87 No. C4 (1982):
                       3052–60.

Ho   Ho Peng Yoke. "Ancient and mediaeval observations of
                       comets and novae in Chinese sources." *Vistas in
                       Astronomy* 5 (1962): 127–225.

Houlden-Stephenson   Houlden, Michael A., and F. Richard Stephenson. *A
                       Supplement to the Tuckerman Tables.* Vol. 170 of
                       APS Memoirs. Philadelphia: American
                       Philosophical Society, 1986.

Housman   Housman, A. E. "Manilius, Augustus, Tiberius,
                       Capricornus, and Libra." *CQ* 7 (1913): 109–14.
                       Reprint in *Classical Papers.* Eds. James Diggle and
                       F. Goodyear. Vol. II. Cambridge: Cambridge Univ.
                       Press, 1972: 867–72.

Hunger et al.   Hunger, Hermann, F. Richard Stephenson, Christopher
                       Walker, and Kevin Yau. *Halley's Comet in History.*
                       London: British Museum Publications, 1985.

Hughes

Hughes, David W. "Cometary outbursts, a brief survey." *Quarterly Journal of the Royal Astronomical Society* 16 (1975): 410–27.

Imhoof-Blumer

Imhoof-Blumer, Friedrich. "Die Kupferprägung des mithradatischen Reiches und andere Münzen des Pontos und Paphlagoniens." *Numismatische Zeitschrift 45* (1912): 169–92 + 2 plates.

Jacoby

Jacoby, Felix. *Die Fragmente der Griechischen Historiker,* II.1 text of Nicolaus of Damascus (No. 90): II.2 commentary. Leiden: Brill, 1961, 1963.

Jaeschke et al.

Jaeschke, W., H. Berresheim, and H.-W. Georgii. "Sulfur emissions from Mt. Etna." *Journal of Geophysical Research* 87 No. C9 (1982): 7253–61.

Jansen

Jansen, Jon. "Additions and correctiions to the I. Hasegawa Catalogue." *Vistas in Astronomy* 34 (1991): 179–86.

Kiang (1972)

Kiang, Tao. "The past orbit of Halley's comet." Royal Astronomical Society, London. *Memoirs* 76 (1972): 27–66.

Kiang (1984)

———. "Notes on traditional Chinese astronomy." *Observatory* 104 (1984): 19–23.

Koch

Koch, C. "Untersuchungen zur Geschichte der römischen Venus-Verehrung." *Hermes* 83 (1955): 1–51.

Koch, *RE*

———. "Venus (1)." *RE 8*A.1. (1955): 828–87.

Koenen, *ZPE* 2

Koenen, Ludwig. "Die Prophezeiungen des 'Töpfers'." *ZPE* 2 (1968): 178–209.

Koenen, *ZPE* 3

———. "Nachträge zu Band 1 und 2." *ZPE* 3 (1968): 137–38.

Koenen, *ZPE* 13

———. "Bemerkungen zum Text des Töperorakles und zu dem Akaziensymbol." *ZPE* 13 (1974): 313–19.

Koenen (1984)

———. "A supplementary note on the date of the Oracle of the Potter." *ZPE* 54 (1984): 9–13.

Kraft

Kraft, K. "Zum Capricorn auf den Münzen des Augustus." *JNG* 17 (1967): 17–27.

Kresák

Kresák, L. "The outbursts of periodic comet Tuttle-Giacobini-Kresák." *Bull. Astr. Inst. Czech.* 25 (1974): 293–304.

Kronk   Kronk, G. *Comets: A Descriptive Catalog.* Hillside:
                         Enslow, 1984.
Lachmann   Lachmann, Karl. *Gromatici Veteres.* Vol. 1 of *Die*
                         Schriften der romischen Feldmesser. Berlin: Reimer,
                         1848.
LaMarche-Hirschboeck LaMarche, V., and K. Hirschboeck. "Frost rings in trees
                         as records of major volcanic eruptions." *Nature* 307
                         (1984): 121–26.
Levi   Levi, Mario Attilio. *Ottaviano Capoparte.* 2 vols.
                         Florence: "La Nuova Italia." 1933.
Levy   Levy, David H. *The Quest for Comets. An Explosive*
                         Trail of Beauty and Danger. New York: Plenum
                         Press, 1994.
Linderski   Linderski, Jerzy. "Cicero and Roman Divination." *PP*
                         37 (1982): 12–38.
Loewe   Loewe, Michael. "The Han view of comets." *Bull. of*
                         the Mus. of Far Eastern Antiquities 52 (1980):
                         1–31.
Lugli   Lugli, G. *Fontes Topographiae Romanae.* Vol. 8.
                         Rome: Università di Roma, Instituto di Topographia
                         Antica, 1962.
Machiavelli   Machiavelli, N. *Discorsi.* In *Tutte le opere di Niccolò*
                         Machiavelli, edited by Francesco Flora and Carlo
                         Cordiè. Milan 1949: Vol. I.
Maeyama   Maeyama Yasukatsu. "The oldest star catalogue of
                         China, Shih Shen's Hsing Ching." In *Prismata:*
                         Naturwissenschaftgeschichtliche Studien, Festschrift
                         für Willy Hartner, edited by Maeyama Yasukatsu
                         and W. G. Salzer. Weisbaden: Franz Steiner, 1977:
                         211–46.
Malcovati   Malcovati, E., ed. *Caesaris Augusti Imperatoris*
                         Operum Fragmenta. 4th ed. Turin: Paravia, 1962.
Marquardt   Marquardt, Karl J. *Römische Staatsverwaltung* III².
                         Leipzig: S. Hirzel, 1885.
Marsden   Marsden, Brian G. *Catalogue of Cometary Orbits.* 8th
                         ed. Cambridge, Mass.: Smithsonian Astrophysical
                         Observatory, 1993.

Marsden-Roemer   Marsden, B. G., and E. Roemer. "Basic information and references." In *Comets,* edited by Laurel L. Wilkening. Tucson: Univ. of Arizona Press, 1982: 727.

Mattingly   Mattingly, Harold. *Coins of Roman Empire in the British Museum* I. 1923. Reprint, London: Trustees of the British Museum, 1965.

McCormick et al.   McCormick, M., L. Thomason, and C. Trepte. "Atmospheric effects of the Mt. Pinatubo eruption." *Nature* 373 (1995): 399–404.

McCuskey   McCuskey, S. W. *Introduction to Celestial Mechanics.* Reading: Addison-Wesley, 1963.

Meech   Meech, K. *IAU Circular* No. 5196. 23 February 1991.

Meinel   Meinel, Aden, and Marjorie Meinel. *Sunsets, Twilights, and Evening Skies.* Cambridge: Cambridge Univ. Press, 1983.

Menzel-Pasachoff   Menzel, Donald H., and Jay M. Pasachoff. *Stars and Planets.* Boston: Houghton Mifflin Co., 1983.

Molnar   Molnar, Michael R. "The Ides of March." *Celator* 8.11 (1994): 6–10.

Mommsen   Mommsen, Theodor. "Das datum der Erscheinung des Kometen nach Caesars Tod." *RBN* 43 (1887): 402–405. Reprint in *Gesammelte Schriften* IV. Berlin: Weidmann, 1906: 180–82.

Morganti-Tomei   Morganti, Giuseppe, and Maria Tomei, "Ancora sulla via Nova." *MEFR* 103.2 (1991): 551–74.

Mucke-Meeus   Mucke, Hermann, and Jean Meeus. *Canon der Sonnenfinsternisse –2003 bis +2526.* Vienna: Astronomisches Buro, 1983.

Mynors   Mynors, R. A. B. *Virgil: Georgics.* Oxford: Oxford Univ. Press, 1990.

Nash   Nash, E. *A Pictorial Dictionary of Ancient Rome.* 2nd ed. I. New York: Praeger, 1968.

Needham   Needham, Joseph, and Wang Ling. *Science and Civilization in China* III. Cambridge: Cambridge Univ. Press, 1959.

Neugebauer   Neugebauer, O. *A History of Ancient Mathematical Astronomy.* 3 vols. Berlin/New York: Springer-Verlag, 1975.

Nisbet

Nisbet-Hubbard

Norden

Ogilvie

Oltramare

Oppolzer

Page

Panciera

Pang et al.

Pascal

Payne-Gaposchkin

Pensabene

Pesce

Petau

Pingré

Pingree

Nisbet, R. G. M. "Virgil's *Fourth Ecolgue:* Easterners and Westerners." *BICS* 25 (1978): 59–78.

Nisbet, R. G. M. and Margaret Hubbard, *A Commentary on Horace: Odes 1.* Oxford: Oxford Univ. Press, 1970.

Norden, Eduard. *Die Geburt des Kindes.* 1924. Reprint, Stuttgart: Teubner, 1958.

Ogilvie, Robert M. *A Commentary on Livy.* Reprint with addenda. Oxford: Oxford Univ. Press: 1978.

Oltramare, Paul. *Sénèque, Questions Naturelles.* 2 vols. Budé ed. Paris: Les Belles Lettres, 1961.

von Oppolzer, Theodor. *Canon of Eclipses.* 1887. Translated by Owen Gingerich. New York: Dover, 1962.

Page, T. E. *P. Vergili Maronis Bucolica et Georgica.* London: Macmillan, 1898.

Panciera, S. *"Olearii."* In *The Seaborne Commerce of Ancient Rome. MAAR* 36 (1980): 235–50.

Pang, K., H.-h. Chou, and D. Pieri. "Climatic impacts of the 44–42 B.C. eruptions of Etna reconstructed from ice core and historical records." *Eos* 67 (1986): 880–81.

Pascal, Carlo. *Studii di Antichità e Mitologia.* Milan: U. Hoepli, 1896.

Payne-Gaposchkin, C. *The Galactic Novae.* Amsterdam: North Holland, 1957.

Pensabene, P. "Scavi nell' area del tempio della Vittoria e del Santuario della Magna Mater sul Palatino." *QArchEtr* 16 (1988): 54–67.

Pesce, G. "Sidus Iulium." *Historia.* Milan-Rome 7 (1933): 402–15.

Petau, Denys. *Opus de doctrina temporum.* 3 vols. Antwerp: Gallet, 1703.

Pingré, Alexandre Guy. *Cométographie ou Traité Historique et Théorique des Comètes.* 2 vols. Paris: Impr. Royale, 1783–84.

Pingree, David. "Petosiris." In *The Dictionary of Scientific Biography.* Vol. 10. New York: Scribners, 1974: 547–49.

Platner-Ashby   Platner, Samuel, and T. Ashby. *A Topographical Dictionary of Ancient Rome.* London: Oxford Univ. Press, H. Milford, 1929.

Press et al.   Press, Willaim H., Brian P. Flannery, Saul A. Teukolsky, and William T. Vetterling. *Numerical Recipes: The Art of Scientific Computing.* Cambridge: Cambridge Univ. Press, 1986.

Rampino et al.   Rampino, Michael R., and Stephen Self, and Richard B. Stothers. "Volcanic winters." *Annual Review of Earth and Planetary Sciences* 16 (1988): 73–99.

Rawson   Rawson, E. "Caesar: Civil war and dictatorship" and "The aftermath of the Ides." *CAH* IX². Cambridge: Cambridge Univ. Press, 1994: 424–67, 468–90.

Rehak   Rehak, Paul. "The Ionic temple relief in the Capitoline: the temple of Victory on the Palatine?" *JRA* 3 (1990): 172–86.

Reif   Reif, F. *Fundamentals of Statistical Mechanics and Thermal Physics.* New York: McGraw-Hill, 1965.

Rettig   Rettig, T. W., S. C. Tegler, D. J. Pasto, and M. J. Mumma. "Comet Outbursts and Polymers of HCN." *The Astrophysical Journal* 398 (1992): 293–298.

Richardson (1)   Richardson, Lawrence, Jr. *New Topographical Dictionary.* Baltimore: Johns Hopkins Univ. Press, 1992.

Richardson (2)   Richardson, Robert S. *Getting Acquainted with Comets.* New York: McGraw-Hill, 1967.

Richter (1949)   Richter, N. "Helligkeitsschwankungen der Kometen und Sonnentätigkeit. II Statistik und Theorie abnormer Lichtausbrüche von Kometenkernen." *Astr. Nach.* 277 (1949): 12–30.

Richter (1963)   Richter, Nicholaus B. *The Nature of Comets.* Translated and revised by Arthur Beer. London: Methuen, 1963.

Ridgway   Ridgway, James D. "Physical considerations concerning the hypothesis that volcanic dust may have obscured the viewing of the supernova of 1054 in Europe." Unpublished honor thesis, College of William and Mary, 1994.

Robock-Matson   Robock, A., and M. Matson. "Circumglobal transport
                    of the El Chichón dust cloud." *Science* 221 (1983):
                    195–97.
Rogers   Rogers, R. S. "The Neronian comets." *TAPA* 84 (1953):
                    237–49.
Rose   Rose, H. J. *The Eclogues of Vergil.* Berkeley: Univ. of
                    California Press, 1942.
Sachs-Hunger   Sachs, Abraham, and Hermann Hunger. *Astronomical
                    Diaries and Related Texts from Babylon.* 2 vols.
                    Vienna: Austrian Academy of Sciences, 1988–89.
Sagan-Druyan   Sagan, Carl, and Ann Druyan, *Comet.* New York:
                    Random House, 1985.
Schanz-Hosius   Schanz, M. *Geschichte der römischen Literatur.* 2 vols.
                    4th rev. ed. by C. Hosius. Munich: Beck,
                    1927–1935.
Schiche   Schiche, Th. "Zu Ciceros Briefen an Atticus." *Hermes*
                    18 (1883): 588–615.
Schilling   Schilling, Robert. *La Religion Romaine de Vénus.* 2nd
                    ed. Paris: Boccard, 1982.
Schippa   Schippa, F. "Una dedica alla Vittoria dalla casa di
                    Augusto al Palatino." *RPAA* 53–54 (1980–82):
                    283–95.
Schmidt   Schmidt, O. E. *Der Briefwechsel der M. Tullius Cicero.*
                    Leipzig: Teubner, 1893.
Schmidt (1883)   ———. "Die Zeit der *lex Antonia Cornelia de
                    permutatione provinciarum* (44 vor Ch.)." *Neue
                    Jahr. f. Philol. u. Pädag.* 127 (1883): 863–65.
Schmitthenner   Schmitthenner, W. *Oktavian und das Testament Cäsars.*
                    2nd ed. Munich: Beck, 1973.
de Schodt   de Schodt, A. "Le *Sidus Iulium* sur des monnaies
                    frappées après la mort de César." *RBN* 43 (1887):
                    329–405.
Schove   Schove, D. Justin, in collaboration with Alan Fletcher.
                    Chronology of Eclipses and Comets A.D. 1–1000.
                    Dover, N. H.: Boydell, 1984.
Schwyzer   Schwyzer, Hans-Rudolf. *Chairemon. Klass.-Phil.
                    Studien* 4: Leipzig 1932.
Scott   Scott, K. "The *Sidus Iulium* and the apotheosis of
                    Caesar." *CPh* 39 (1941): 257–72.

Scuderi [1]   Scuderi, Louis A. "Oriental sunspot observations and
                       volcanism." *Quarterly Journal of the Royal
                       Astronomical Society* 31 (1990): 109–120.

Scuderi [2]   ———. "Tree-ring evidence for climatically effective
                       volcanic eruptions." *Quaternary Research* 34
                       (1990): 67–85.

Scullard   Scullard, H. H. *Festivals and Ceremonies of the Roman
                       Republic.* London: Thames and Hudson, 1981.

Sekanina   Sekanina, Z. "The Problem of split comets in review."
                       In *Comets,* edited by Laurel L. Wilkening. Tucson:
                       Univ. of Arizona Press, 1982): 251–87.

Shackleton Bailey   Shackleton Bailey, D. R. *Cicero's Letters to Atticus.* 7
                       vols. Cambridge: Cambridge Univ. Press, 1965–70.
                       ———. *Cicero: Epistulae ad Familiares.* 2 vols.
                       Cambridge: Cambridge Univ. Press, 1977.
                       ———. *Cicero: Philippics.* Chapel Hill: Univ. of N.
                       Carolina Press, 1986.

Shackleton Bailey,   ———. *Two Studies in Roman Nomenclature.* 2nd ed.
   Two Studies   Atlanta: Scholars Press, 1991.

Simkin-Siebert   Simkin, Tom, and Lee Siebert, with the collaboration of
                       R. Blong, J. Dehn, C. Newhall, R. Pool, and T.
                       Stein. *Volcanoes of the World: A Regional Directory,
                       Gazetteer, and Chronology of Volcanism During the
                       Last 10,000 Years.* 2nd ed. Tucson: Geoscience
                       Press, 1994.

Stephenson (1990)   Stephenson, F. R., "The ancient history of Halley's
                       comet." In *Standing on the shoulders of giants,*
                       edited by Norman Thrower. Berkeley: Univ. of
                       California Press, 1990.

Stephenson (1994)   Stephenson, F. R. "Chinese and Korean star maps." In
                       History of Cartography II.2. Eds. J.B. Harley and D.
                       Woodward, 511–78.

Stephenson et al.   Stephenson, F. R., K. K. Yau, and H. Hunger. "Records
                       of Halley's Comet on Babylonian tablets." *Nature*
                       314 (1985): 587–92.

Stephenson-Foley   Stephenson, F. R., and Neasa Foley. Paper on solar
                       eclipse records in *HS* 27. In preparation.

Stommel   Stommel, H., and E. *Volcano Weather.* Newport, R.I.:
                       Seven Seas Press, 1983.

Stothers-Rampino   Stothers, Richard B., and Michael R. Rampino.
                     "Volcanic eruptions in the Mediterranean before
                     A.D. 630 from written and archaeological sources."
                     Journal of Geophysical Research 88 No. B8.
                     (1983): 6357–71.

Sumner   Sumner, G. V. "Suetonius *Divus Iulius* 86.2 and 88.2:
                     Two Notes." *CPh* 68 (1973): 291–92.

Sutherland   Sutherland, C. H. V. *The Roman Imperial Coinage* I².
                     London: Spink & Son, 1984.

Syme   Syme, R. *Roman Revolution.* 1938. Reprint, Oxford:
                     Oxford Univ. Press, 1960.

Symons   Symons, G. J., ed. *The Eruption of Krakatoa and
                     Subsequent Phenomena.* London: Trubner, 1888.

Tamura   Tamura, Sennosuke. *Tôyôyin no Kagaku to Gijutsu*
                     [Essays in the History of East-Asian Science and
                     Technology]. Tokyo 1958.

Taylor   Taylor, Lily Ross. "On the chronology of Cicero's
                     letters to Atticus, Book XIII." *CPh* 32 (1937):
                     228–40.

Taylor, *Divinity*   ———. *The Divinity of the Roman Emperor.* 1931.
                     Reprint, Atlanta: Scholars Press, n.d.

Thiele   Thiele, Georg. *Antike Himmelsbilder.* Berlin:
                     Weidmann, 1898.

Thomas   Thomas, Richard F. *Virgil, Georgics I–II.* Cambridge:
                     Cambridge Univ. Press, 1988.

Tung   Tung Tso-Pin. *Chung-kuo nien-li tsung-p'u.*
                     [*Chronological tables of Chinese history* I.] Hong
                     Kong: Hong Kong Univ. Press, 1960.

Tyrrell-Purser   Tyrrell, Robert, and Louis Purser. *Correspondence of
                     Cicero,* V². 1915. Reprint, Hildesheim: Olms, 1969.

Ulrich   Ulrich, Roger B. "Julius Caesar and the creation of the
                     Forum Iulium." *AJA* 97 (1993): 49–80.

Ville   Ville, Georges. *La Gladiature en Occident des Origines
                     à la Mort de Domitien.* Rome: École Française de
                     Rome, 1981.

Volkmann   Volkmann, Hans. *Res Gestae Divi Augusti.* 3rd ed.
                     Berlin: De Gruyter, 1969.

Vollenweider   Vollenweider, Marie-Louise. *Die Porträtgemmen der römischen Republik.* 2 vols. Mainz: Zabern, 1972–74.

Vsekhsvyatskii   Vsekhsvyatskii, Sergey K. *The Physical Characteristics of Comets.* Translated and edited by the Israel Program for Scientific Translations, with the support of NASA and the NSF. Jerusalem: Monson, 1964.

Wagenvoort   Wagenvoort, H. "Virgil's *Fourth Eclogue* and the *sidus Julium.*" *Meded. Kon. Ned. Akad. v. Wetensch. afd Lett.* 67A.1. 1929. Reprinted in *Studies in Roman Literature, Culture and Religion.* Leiden: Brill, 1956: 1–29.

Wagner   Wagner, Philippus (G. P. E.) *Virgil.* breviter enarravit. 3rd ed. Leipzig: Teubner, 1861.

Weber   Weber, W. *Princeps: Studien zur Geschichte des Augustus* I. Stuttgart: W. Kohlhammer, 1936.

Weinstock   Weinstock, Stefan. *Divus Julius.* Oxford: Oxford Univ. Press, 1971.

Weinstock (1957)   ———. "Victoria and *invictus.*" *HTR* 50 (1957): 211–47.

Weiss   Weiss, P. "Die 'Säkularspiele' der Republik—Eine annalistische Fiktion?" *MDAI (R)* 80 (1973): 205–17.

West   West, David. "On serial narration and on the Julian star." *Proceedings of the Virgil Society* 21 (1993): 1–16.

Whiston   Whiston, William. *A New Theory of the Earth.* 4th ed. rev. London: Tooke & Motte, 1725.

Williams (1871)   Williams, John. *Observations of comets, from B.C. 611 to A.D. 1640. Extracted from the Chinese annals.* London: Strangeways and Walden, 1871.

Williams   Williams, R. D. *Virgil, Aeneid V.* Oxford: Oxford Univ. Press, 1960.

Wilson et al.   Wilson, C., N. Ambraseys, J. Bradley, and G. Walker. "A new date for the Taupo eruption, New Zealand." *Nature* 288 (1980): 252–53.

Wiseman   Wiseman, T. P. "The temple of Victory on the Palatine." *AntJ* 61 (1981): 35–52.

Wissowa   Wissowa, G. *Religion und Kultus der Römer.* 2nd ed.
                   1912. Reprint, Munich: Beck, 1971.

Wolfram   Wolfram, E. "The political function of astronomy and
                   astronomers in Han China." In *Chinese Thought and
                   Institutions,* edited by J. K. Fairbank. Chicago:
                   Univ. of Chicago Press, 1957: 33–70.

Wu-Liu   Wu Shouxian, and Liu Ciyuan. "Ancient Chinese
                   astronomical observations related to the stellar
                   background on the sky." *Publ. Shanxi Observatory*
                   13.2 (1990): 31–38.

Wunderlich   Wunderlich, E. C. G. and C. G. Heyne, *P. Virgilii
                   Maronis Opera.* Vol. 1. Leipzig: Hahn, 1822.

Yavetz (1969)   Yavetz, Zvi. *Plebs and Princeps.* Oxford: Oxford Univ.
                   Press, 1969.

Yavetz (1984)   _____. "The *Res Gestae* and Augustus' public
                   image." In *Caesar Augustus: Seven aspects.* Edited
                   by F. Millar and E. Segal. Oxford: Oxford Univ.
                   Press, 1984: 1–36.

Yeomans   Yeomans, Donald. *Comets: A Chronological History of
                   Observation, Science, Myth, and Folklore.* New
                   York: John Wiley & Sons, 1991.

Yeomans (2)   _____."Comets, historical apparitions." In *The
                   Astronomy and Astrophysics Encyclopedia.* Edited
                   by Stephen Maran. New York: Van Nostrand
                   Reinhold, 1992: 114–18.

Zanker   Zanker, P. *The Power of Images in the Age of Augustus.*
                   Ann Arbor: Univ. of Michigan Press, 1988.

Zhuang-Wang   Zhuang Weifeng and Wang Lixing, *Zhongguo Gudai
                   Tianxiang Jilu Zongji. [Complete collection of
                   records of heavenly phenomena.]* Nanjing: Tiansu,
                   1988: 948–52.

Zielinski   Zielinski, G. "Stratospheric loading and optical depth
                   estimates of explosive volcanism over the last 2100
                   years derived from the Greenland Ice Sheet Project
                   2 ice core." *Journal of Geophysical Research* 100
                   No. D10 (1995): 20937–55.

Zielinski et al.   Zielinski G., P. Mayewski, L. Meeker, S. Whitlow, M.
                           Twickler, M. Morrison, D. Meese, A. Gow, and R.
                           Alley. "Record of volcanism since 7000 B.C. from
                           the GISP2 Greenland ice core and implications for
                           the volcano-climate system." *Science* 264 (1994):
                           948–52.

Ziolkowsi (1989)   Ziolkowsi, Adam. "The Sacra Via and the temple of
                           Iuppiter Stator" *Opuscula Romana* XVII: 17 (1989):
                           225–39.

Ziólkowsi (1992)   _____. *The Temples of Mid-republican Rome and
                           their Historical and Topographical Context.* Rome:
                           L' Erma di Bretschneider, 1992.

INDEX I

NAMES AND SUBJECTS

217

INDEX II
GREEK WORDS

INDEX III

SOURCES

[ ] enclose works falsely, or doubtfully, attributed to an author.

(A) LITERARY

Amm. Marc. (Ammianus Marcellinus)
25.2.7	140.20
25.10.1-2	140.20

Appian, Bella Civilia
2.23.88	46
2.68.281	22.5
2.102.424	22.5
2.106.440	34.25; 34.26
2.106.442	35.27
2.106.443	24
2.147.613	47.24
2.148.616-617	56
3.2.3-3.9	2.3
3.9.32	50.33
3.14.49	1
3.17.62-64	10.24
3.20.73-75	10.24
3.21.77-23.89	10.24
3.23.87	45; 45.13
3.24.90-91	46
3.28.105-106	2.3
3.28.107	
(Text 15, p. 168)	2.3; 52
3.29.111-115	2.4
3.39.156	2.4
4.4.14	100.1
4.4.15	96.4; 141.24

Aristot. (Aristotle)
Mete. (Meteorologica)
1.6.343b	87.66; 161.4
1.7.344a 23-24	139.17

[*Mund.*] *(De Mundo)*
395b11	93.74

Ascon. (Asconius)
13C.15	186.8; 187.9
32-33C	47.24

Augustus
RG (Res Gestae)
7.3	51
22.1	48.25
22.2	51; 51.35; 57.55

Vit. (De vita sua)
fr. 6
(Text 1, p. 158)	3.6; 50.32; 84; 88; 136; 136.7; 144; 144.33; 147; 152

Avienius
ap. Serv. *Aen.*
10.272
(Text 13, p. 168)	68; 139.14; 145; 145.33; 146

Aratea 1814-19   135.4

Caes. (Caesar)
B.G. (Bellum Gallicum)
1.16.2	86

[*BAfr.*] *(Bellum Africum)*
98	183

De astris 98.12

Calp. Sic. (Calpurnius Siculus)
Ecl. (Eclogae)
1.77-83
(Text 33, p. 176)	136.6; 144;
1.80	106.41
1.81	93.75

(B) NON-LITERARY